Serial Networked
Field Instrumentation

WILEY SERIES IN MEASUREMENT SCIENCE AND TECHNOLOGY

Chief Editor

Peter H. Sydenham
Australian Centre for Test & Evaluation
University of South Australia

**Instruments and Experiences:
Papers on Measurement and Instrument Design**
R. V. Jones

Handbook of Measurement Science, Volume 1
Edited by P. H. Sydenham

Handbook of Measurement Science, Volume 2
Edited by P. H. Sydenham

Handbook of Measurement Science, Volume 3
Edited by P. H. Sydenham and R. Thorn

Introduction to Measurement Science & Engineering
P. H. Sydenham, N. H. Hancock and R. Thorn

Temperature Measurement
L. Michalski, K. Eckersdorf and J. McGhee

Technology of Electrical Measurements
Edited by L. Schnell

**Traceable Temperatures: An Introduction to
Temperature Measurement and Calibration**
J. V. Nicholas and D. R. White

Image Based Measurement Systems
F. van der Heijden

Serial Networked Field Instrumentation
J. R. Jordan

Serial Networked Field Instrumentation

J. R. Jordan
Department of Electrical Engineering
University of Edinburgh, UK

JOHN WILEY & SONS
Chichester • New York • Brisbane • Toronto • Singapore

Copyright © 1995 by John Wiley & Sons Ltd,
Baffins Lane, Chichester,
West Sussex PO19 IUD, England

All rights reserved.

No part of this book may be reproduced by any means,
or transmitted, or translated into a machine language
without the written permission of the publisher.

Other Wiley Editorial Offices

John Wiley & Sons, Inc., 605 Third Avenue,
New York, NY 10158-0012, USA

Jacaranda Wiley Ltd, G.P.O. Box 859,
Brisbane, Queensland 4001, Australia

John Wiley & Sons (Canada) Ltd, 22 Worcester Road,
Rexdale, Ontario M9W 1L1, Canada

John Wiley & Sons (SEA) Pte Ltd, 37 Jalan Pemimpin #05-04,
Block B, Union Industrial Building, Singapore 2057

Library of Congress Cataloging-in-Publication Data
Jordan, J. R.
 Serial networked field instrumentation / J. R. Jordan.
   p. cm. — (Wiley series in measurement science and
 technology)
  Includes bibliographical references and index.
  ISBN 0 471-95326 1: $60.00
  1. Digital control systems. 2. Engineering instruments.
 3. Telecommunication systems. I. Title. II. Series.
TJ223.M33J67 1995   95-934
629.8 — dc20   CIP

British Library Cataloguing in Publication Data

A catalogue record for this book is available from the British Library

ISBN 0 471 953261

Typeset in $10\frac{1}{2}/12\frac{1}{2}$ Times from author's disks by Laser Words, Madras, India
Printed and bound in Great Britain by Biddles Ltd, Guildford, Surrey

Contents

PREFACE ix

SERIES EDITIOR'S PREFACE xi

1 INTRODUCTION 1

2 COMMUNICATION MEDIA AND CODES 17
- 2.1 Introduction 17
- 2.2 The twisted pair and coaxial cable 17
- 2.3 Fibre optic networks 20
- 2.4 The radio link 27
- 2.5 4–20 mA analogue signalling and intrinsic safety 32
- 2.6 Digital codes 35
- 2.7 Error detection and correction 36
- 2.8 Spread spectrum encoding 38
- 2.9 Power line communication networks 39
- 2.10 The EIA RS standards 42

3 STANDARDS 45
- 3.1 Introduction 45
- 3.2 The Man-Machine and Machine-Machine Interface 45
- 3.3 Media Access Methods 48
- 3.4 Standards 52
- 3.5 Standards Organizations 54
- 3.6 The Standard Document 57

4 FACTORY AUTOMATION AND PROCESS CONTROL 61
- 4.1 Introduction 61
- 4.2 Process Control 62
- 4.3 Factory Automation 64
- 4.4 The Fieldbus 68
- 4.5 Intelligent Transducers 76
- 4.6 Power over the Bus 81
- 4.7 Safe Systems 83

5 LABORATORY AND MEDICAL AUTOMATION 87
- 5.1 Introduction 87
- 5.2 Laboratory Instrumentation 87
- 5.3 Medical Instrumentation 91

6 INTELLIGENT BUILDINGS 97
- 6.1 Introduction 97
- 6.2 Domestic Buildings 97
- 6.3 Commercial Buildings 104
- 6.4 Standards 107

7 TRANSPORT 109
- 7.1 Introduction 109
- 7.2 Military Systems 109
- 7.3 Civil Aviation 113
- 7.4 Vehicles 113
- 7.5 Road Automation 118

8 ELECTRONIC SYSTEMS 121
- 8.1 Introduction 121
- 8.2 The Transputer 121
- 8.3 Inter-Integrated Circuit (I^2C) bus 125

9 THE CONNECTED FUTURE 131
- 9.1 Serial Networks 131
- 9.2 Standards and Products 133
- 9.3 Systems Integration 135
- 9.4 Flexible Manufacturing 136
- 9.5 Conclusions 138

APPENDIX A STANDARDS 141
- A.1 Introduction 141
- A.2 ARCNET 141
- A.3 Military standards 1553B and 1773 145
- A.4 IEEE 1118 150
- A.5 ARINC 629 152
- A.6 PROFIBUS 154
- A.7 FIP: The Factory Instrumentation Protocol 157
- A.8 HART 159
- A.9 CEbus and the Home Bus System 162
- A.10 Echelon LonWorks 165
- A.11 IEEE P1073 168
- A.12 Controller Area Network, CAN 170
- A.13 IEC Fieldbus 174

APPENDIX B STANDARDS ORGANISATIONS 177

APPENDIX C INTEGRATED CIRCUITS 179
- C.1 ARCNET – COM20051 179
- C.2 CAN – PCA82C200 193
- C.3 Echelon Neuron Chip 210

REFERENCES 227

INDEX 237

Preface

This book has its origins in a series of research projects investigating low-power instrumentation for serial networked measurement and control systems. This work revealed a large activity concerned with establishing standards that would enable spatially separated devices to interwork successfully using a common communication interface. Initially standards arose from and were applied to specific market sectors. The dramatic increase in the capability of integrated circuits has facilitated the development of relatively low-cost devices that satisfy the communication requirements of a wide range of market areas. Consequently standards (proprietary, national and international) are appearing that now apply to a wide range of market sectors. Twelve serial network communication standards are specifically discussed in this book; many more exist and they continue to be developed.

The apocryphal, comment, 'of course we believe in standards, that's why we have so many of them' has a hollow ring, especially when it is remembered that difficulties associated with creating multi-vendor networks are one of the most irritating features of modern information technology. This book presents an up-to-date overview of serial networking technology for measurement and control applications. Hopefully it will provide a frame of reference for anyone contemplating the development of yet another standard.

Diane Armstrong and Caroline Saunders deserve my special thanks for their patience in preparing the manuscript and solving all of my secretarial problems. Wang Jian-Zhou, who comes from Beijing, in the People's Republic of China, and is researching fuzzy logic control algorithms with me, is thanked for the care with which he converted rough sketches for diagrams into a form suitable for computer storage and manipulation. I am indebted to my editor, Ann-Marie Halligan.

Series Editor's Preface

The Series provides authoritative books, written by internationally-acclaimed experts, on the topics which cover the fundamental science and engineering practice of measurement and sensing.

Measurement Systems will often involve many different measurements made with sensors providing signals bearing information on electronic communication networks. Such networks are often large and complex, involving thousands of sensors.

This important work provides a thorough reference source on the science and practice of arranging sensor communication with decision making nodes–people or computers. Signals will usually be sent back from these nodes to bring about control of the system part.

Peter Sydenham
Editor in Chief

1
Introduction

In all market sectors where products using measurement and control instrumentation can be found, the cost (hardware, software, installation and commissioning) of field-located devices is a significant proportion of the total cost of the system. Conventional wiring uses a star (or stem and fan) connection topology, as shown in Figure 1.1. A large part of this cost is due to the wiring used to interconnect field devices. A serial connection method, using a multi-drop bus as shown in Figure 1.2, considerably simplifies the wiring problem. Several industrial studies have shown that savings of at least 25% can be expected when serial digital bus techniques are adopted.

When devices are connected by a ring or bus connection some form of communications protocol is required to ensure that transmissions to and from devices can be completed without error. It is highly desirable that the protocol used conforms to a widely accepted standard. The experiences of the information technology industry indicate that users will want and expect equipment purchased to operate in an open system environment. It can be expected that interoperability and interchangeability of field devices will become a major design objective.

The digital link to the field device will encourage the incorporation of higher levels of functionality in field devices. The increasing capability and decreasing cost of integrated circuit components indicates that this will be economically feasible. Indeed once devices are available to operate on a standard bus, product differentiation will be achieved by increasing functionality by means of integrated circuit technology. Typical functions offered by a modern digital field device will include: some degree of self-calibration, validation of sensor outputs, programmable (via the bus) operating modes and built-in fault detection. A serial digital device will not require calibration over a specific measurement range (e.g. to fit the 4–20 mA signalling range used in conventional instrumentation). One result of this will be the elimination of the need to use more than one measurement system when the variable to be monitored covers a very wide range. Installation of devices will require a minimum level of field operator involvement.

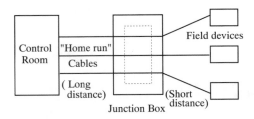

Figure 1.1
Conventional field device wiring topology

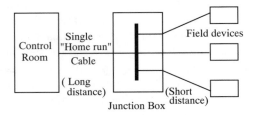

Figure 1.2
Field device wiring topology based on the use of a multi-drop serial bus

Communication with the device will be by transmissions over the bus and by means of a hand-held portable communication terminal. The hand-held terminal will communicate with the system by a clip-on connection to the bus or by low-power optical, inductive or radio links with the field device. A field device will probably have no field-adjustable controls; all control, adjustment and configuration operations will be under software control. It is expected that a by-product of the serial digital approach to field instrumentation will be a significant reduction in the quantity and complexity of engineering support documentation supplied to the user. Not only will the system intelligence be distributed spatially throughout the system, but also the technical knowledge base will be distributed through the system software (including the field device software system).

It should be noted that serial digital connection techniques were discussed as soon as the first digital computers were used to control an industrial process. Collins (1968) presents an early survey of industrial measurement techniques for on-line computers.

Developments in integrated circuit technology, software and communication networks, have only just reached the stage where low-cost complex devices can be manufactured for field applications. Progress towards an internationally agreed approach to the application and user interface with the field device is still very slow.

1 INTRODUCTION

While the serial interconnected devices will be relevant for small- as well as large-scale measurement and control systems, the value of the serial digital approach to this instrumentation can only be completely appreciated by considering the typical large-scale industrial application. The BNFL thermal oxide reprocessing plant (THORP) is a good example of a large-scale plant. (this plant is located at Sellafield in the UK). It effectively controls a continuous process in which spent fuel is dissolved in a liquid and fed through a series of chemical operations designed to remove solid particles from the liquid. Radioactive waste is isolated and the resulting elements, such as uranium, are purified. In addition to the main continuous process, batch processes operating intermittently during a 24 hour period have to be monitored and controlled. Taylor Instrument (now part of ABB) installed six of their MOD 300 systems (first introduced in the USA in 1984) to control this plant. The MOD 300 systems communicate over a Taylor Instrument proprietary (open-system interconnect-compatible), 1 Mb/s, token passing bus using high grade co-axial cable configured in a ring topology. Six rings are used with dual (separately trayed) connections to provide a dual active system to give enhanced reliability. The six rings interconnect to form a site-wide network for process management and safety analyses. A total of 5000 analogue process inputs, 12 000 digital inputs, 6000 digital output and over 600 control loops are required to operate this plant. An additional 700 measurements are made to satisfy environmental monitoring requirements. Clearly the field wiring overhead for a project of this magnitude is considerable. THORP was put on-line in 1994 a decade after the MOD 300 technology was first introduced in the USA.

Standards have become increasingly important in measurement and control over the past decade. A very large international activity has driven standards development to the point where a detailed knowledge of standard documentation is seen to be an essential pre-requisite of product development. In fact product definition and standard development are now commonly seen as parallel activities, and standards can appear before products reach the market place. The standard document is based on the work of an expert committee. The combined expert knowledge of each committee is distilled into a standards document that defines a concept that industry is expected to interpret to enable products to be manufactured in conformity with the standard. Hardware and implementation details are not normally specified by a standard. Hence, testing for conformity of a product to a particular standard is an important aspect of product development. A wide range of standards affect the development of measurement and control products. In this book attention is restricted to serial bus standards.

It is unfortunate that proliferation of serial bus standards appears to be the norm. Dorey (1977) noted that there were at that time, at least five international organizations known to be preparing at least 22 interface standards, while the number of national and commercial standards in preparation outnumbers the international standards. This book discusses more than 18 standards designed for measurement and control applications.

The interconnection and interworking of multivendor systems has been a major concern of engineers developing data processing and data collecting equipment. A need for a standardized communication architecture was recognized internationally in 1977 when the first attempts were made to produce a general description of the communication process. It is interesting to note the way that communication standards have developed with respect to the evolution of integrated circuit technology. Table 1.1 shows the chronological development of measurement and control communication standards.

The open system interconnect (OSI) model comprehensively defines the tasks needed to establish an interconnection, and subdivides these tasks into seven interfaced layers, with the bottom layer defining physical details and the top layer defining notation and operations specific to the particular application. It is important to note that the model does not define the protocols to be used by each layer. The first definition of the model appeared in 1978 (ISO 7498) but since then many additions have been made and these are now collectively referred to as the Open Systems Interconnect.

The subdivision of the total communication operation into layers is arbitrary but it is a necessary requirement to achieve a flexible easily changed system.

Table 1.1
Chronological development of measurement and control communication standards

1951	Discrete transistors commercially available
1955	4–20 mA process control instrumentation
1960	Small-scale integration (3–100 components per chip)
1969	Large-scale integration (10^3–10^4 components per chip), RS-422/423 serial interface standard
1970	BS 3586, 4–20 mA standard
1975	Ethernet
1977	Arcnet
1978	ISO seven-layer reference model for OSI
1978	MIL-STD-1553B
1980	General Motors factory automation working group
1981	Smart transducers
1983	ISO 7498-international standard for the basic reference model for OSI
1984	MAP version 1.0
1985	IEC TC 65C/WG6 — fieldbus working group
1985	IEEE 802 series of standards
1986	UK DTI CIMAP 1553B demonstration field bus at the National Exhibition Centre, Birmingham, England
1987	MAP version 3.0
1994	Fieldbus standard?

1 INTRODUCTION

Each layer is designed to be independent of the layers above and below it, so layers may be changed as technological capabilities change without affecting the operation of the overall system. Some of the specified seven layers will not be needed at all in some applications and some layers may be divided into sublayers.

A common criticism of the OSI is that it is an example of a standard with too many options. It has led users to pick subsets which suit their needs. This criticism can be extended to the IEC Fieldbus Standard (discussed later in this book) which is firmly based on the OSI Communication model.

Serial digital communication methods have been used extensively to link computing systems. The serial networks discussed in this book are based on this work. The bit-oriented protocols, HDLC/SDLC, have been widely used. HDLC is now an ISO standard (ISO 3309). Ethernet, which first appeared in 1975, is now an IEEE/ISO standard (Smythe, 1993). The IEEE 802 series of standards has made a considerable impact on digital communication technology. The integrated circuit implementation of those standards which is now available has considerably reduced the cost of implementing these serial networks.

The high-level data link control (HDLC) uses synchronous framed transmissions which include a 16 or 32 bit frame check sequence for error control purposes. A command/response protocol controls access to the bus. Some stations on the bus are defined only to respond to commands. The synchronous data link control, SDLC, (a widely used IBM proprietary protocol) is essentially a subset of HDLC.

The Institute of Electrical and Electronic Engineers (IEEE) is an important standardization body in the USA. Its 802 series of standards is based on a three-layer architecture for local network access and control which corresponds with the lower two layers (physical connection and bus access control) of the open system interconnect communication model. The top two layers of the 802 communication model (media access control (MAC) and logical link control (LLC)) corresponds to the data link layer of the OSI model. The LLC layer (defined by IEEE Standard 802.2) is common to the four MAC and physical layers defined by the series of standards; it provides for connection-oriented and connectionless data transfer, multipoint and broadcast modes of operation. The series of standards covers CSMA/CD (carrier sense multiple access/with collision detection) and token passing access control methods. The four standards are:

- IEEE 802.3 CSMA/CD medium access control (baseband coaxial, 10 Mb/s; unshielded twisted pair, 1/10 Mb/s; broadband coaxial 10 Mb/s)
- IEEE 802.4 token-bus medium access control (broadband coaxial, 1/5/10 Mb/s; Carrierband, 1/5/10 Mb/s; Optical fibre, 5/10/20 Mb/s)

- IEEE 802.5 token-ring medium access control (shielded twisted pair, 1/4 Mb/s)
- fibre distributed data interface (FDDI) token-ring medium access control (optical fibre 100 Mb/s).

A feature of these LAN standards is the number of options allowed at each layer. The large size of the IEEE Standards documentation established a trend which the IEC Fieldbus standard is continuing.

This book was written as a contribution to the development of a universal approach to serial field networks. The large number of currently available networks are all driven by a significant sector of the measurement and control market so it is probably unrealistic to expect the acceptance of just one standard. Nevertheless the use of scarce, expert, human resources to develop yet another network should surely be constrained and deployed to create products that would make full use of a truly international serial network.

The trend is towards equipment designed for a particular market sector to adopt networks developed from work in other market sectors. Arcnet and CAN illustrate this trend. Arcnet was originally developed for use in office applications requiring the interconnection of computing equipment, but is now being widely used in industrial automation applications of measurement and control. CAN was devised for applications in vehicle systems and is consequently able to deal with a range of real-time conditions and responses. Its short message lengths and speed of response is being utilized in industrial automation applications. Both of these standard networks are supported by integrated circuit components that simplify their use in a wider range of applications than was originally imagined.

The availability of integrated circuit network components is leading instrumentation manufacturers to develop a multinetwork capability. Market differentiation is obtained by appropriately designing the application layer and user interface. Some degree of commonality may be achievable at the physical layer and data link layer of the communication model by using specially designed multifunction integrated circuits.

One intention of this book is to provide sufficient information for the reader to make an informed initial assessment of the suitability of a network for a particular measurement and control application. Chapters 4–8 are organized to present the distinct market sectors where serial networks are used in measurement and control applications. Appendix A presents detailed information on the most widely used networks. Each of the market and application oriented chapters introduces the practical use of these networks. A particular feature of this book is the inclusion of sections to discuss the impact of standards on the development of serial networks and their practical use in products. Standards development and application are an important aspect of the modern product development cycle.

1 INTRODUCTION

Several books present information on digital communication methods and systems. *Data and Computer Communications* by Stallings (1991) is a good example. These books will provide the detailed information on digital communication and the OSI communication model that, if included in this book, would have made it over long. An attempt has been made to cover some of this material in Chapter 2, which includes communication-related information such as the digital codes and error checking methods used in serial networks.

This book has been written from the point of view of the measurement and control engineer working as the user or the supplier of electronic instrumentation.

The audience for this book is expected to be:

- engineers working in measurement and control who wish to obtain an overview of serial networked field instrumentation
- final year undergraduates who wish to obtain an appreciation of the world activity in the serial networks area
- researchers and designers working in the area of serial networked instrumentation and distributed intelligent control who wish to relate their work to product and market oriented world serial networks activity driven by commercial standards.

Hopefully, sufficient information has been presented to enable network selection exercises to be completed and to allow networked systems to be designed without repeating the current body of work.

Chapter 3 presents a discussion of serial network standards and emphasizes the importance of the standardization process for product development. Users want to establish systems based on serial networked instrumentation which will allow the free interchange of equipment from different manufacturers. Clearly the man–machine and machine–machine interface will require to be defined by a well established and widely accepted standard. Media access control is a major feature of this type of standard. The performance of a serial bus system is determined to a considerable extent by the way a device connected to the bus gains control of the media. The major techniques: master–slave, token passing and carrier sense with collision detection (or avoidance) are discussed in Chapter 3. The conflicting requirement of deterministic operation via a master–slave or token passing protocol versus the ease of configuration of masterless systems have been difficult to resolve during the development of the IEC Fieldbus standard. The media access protocol is embedded in the OSI communication model, and this has played an important part in the development of standard serial networks. The manufacturing automation protocol (MAP) was an early user of the OSI model. For field device network applications a simplified version of this model is used when a fast response system is required. A desire to configure intelligent instrumentation with serial network architectures has been a strong motivation for the development of the serial data highway. The design of the user interface for this type of system has a significant impact on the success of the system, and the standardization of the interface is as important as the standardization of the machine–machine serial bus interface.

Standards and standards organization are discussed in Chapter 3. It was felt that this was an essential addition to the core material of this book because of the influence that standards exert on product development; serial networked field instrumentation appears to be particularly strongly affected by standards development. The worldwide standards effort is considerable, with the major effort led by the International Standards Organization and the International Electrotechnical Commission, supported by national bodies such as the British Standards Institute, Association Française de Normalisation, Deutshes Institut für Normung and the American National Standards Institute, coordinating the activities of bodies such as the Institute of Electrical and Electronic Engineers and the Instrument Society of America. From the point of view of the designer of serial networked instrumentation the important standard considerations include the following: should a standard be pre- or post-product, and is the standard document designed to be easy to use to implement products in conformance with the standard?

In the information technology and communications technology areas it has become common practice for standards to be developed in parallel with the development of new products. These pre-product standards appear to have a much greater commercial significance than post-product standards. Post-product standards are based on tried and tested designs which are already impacting a significant market sector. In theory users should be in a position to have a greater influence on pre-product standards. However, in practice users have a low involvement with standards committees and it will therefore still be the market place that determines the success of new products. Since a standard does not provide direct implementation guidance, a manufacturer is always in a position to achieve a successful product by good design and manufacturing procedures.

Networking standards are often long complex documents. In contrast the 4–20 mA analogue signalling standard (BS 3586), which was a post-product standard, required only a few pages. Nevertheless a wide range of products and systems have been based on this standard. Clearly options should be avoided and some form of formal method should be used to define complex operating procedures. It is interesting to note that the IEC Fieldbus standard will be a multivolume document which does not use formal specification methods, but includes a range of options that need to be included before a consensus could be obtained to enable the standard to be completed.

1 INTRODUCTION

Chapter 4 addresses the factory automation and process control application areas. It includes a discussion of the related topics of safe systems, intelligent transducers and a discussion of the concepts that have contributed to the development of the IEC Fieldbus standard.

Process automation is concerned with the automatic operation of continuous processes, such as those commonly found in the petrochemical industry. Analogue 4–20 mA signalling over twisted pairs has been used since the mid 1950s to link control equipment to field-located sensors and actuators. The standard that defines this connection method (BS 3586) was published in 1970, illustrating the long delay characteristic of post-product standardization. A variety of proprietary serial networks were in use in the process control industry in the 1980s, but it was not until integrated circuit technology allowed the low-cost implementation of complex bus protocols that this work could be extended to field devices. National standards activity in Germany and France has progressed to the stage where integrated circuit components are available for use in practical systems. The IEC Fieldbus standard, designed for process control and factory automation applications, is making slow progress but it will hopefully soon be finished. The 4–20 mA signalling method is still in use; it is clear that the move towards fieldbus-based process control will be a slow evolutionary progression.

It is common practice in process control applications for the lower 4 mA of the 4–20 mA signalling range to be used to supply power to remote devices. Fieldbus systems are being developed to allow a power supply signal to be transmitted with the digital communication signal. AC and DC techniques with voltage and current sources have been developed. In this case low-power operation of field devices is desirable and in some applications where intrinsic safety is required a very low power budget will be available for field devices.

Factory automation involves discrete-parts manufacturing operations with equipment grouped into cells which are often connected to a centrally located supervisory system by means of a LAN. There will of course be overlapping areas of interest with process control; for example, the control of electric drives for speed and position control. General Motors in the USA, an extensive user of factory automation technology, established in 1980 a group to investigate the use of standard serial networks to simplify the configuration and operation of equipment in a multivendor environment. This work led to the creation of the Manufacturing Automation Protocol. An objective of modern automation is to provide flexible solutions to manufacturing problems. Flexibility is obtained by a combination of user-friendly software design with the use of fieldbus techniques to simplify installation and configuration, and facilitate the addition and removal of field devices.

Devices operating with the 4–20 mA analogue signalling method can be designed to include a digital capability that can be communicated with by superimposing a digital signal on the analogue signal. Such devices have been called intelligent (or smart) devices although the level will be very low. The important features of a so-called intelligent device are that it should be able to operate with minimum human interference, have programmable functions and be able to take decisions to maintain its performance by changing its response to varying operating conditions. The interconnection of intelligent field devices by a fieldbus will enable intelligence to be distributed and controlled without the need for a permanently attached host system. Integrated circuit technology will allow the complex electronic systems required to implement these functions to be incorporated in sensors and actuators without a large cost penalty. However other considerations, such as system safety and reliability, may make this approach less attractive. The safety implications of programmable electronic systems and complex software packages are discussed in Chapter 4.

The concepts that have led to the development of the IEC Fieldbus standard are discussed in Chapter 4. The competing standard bus systems that will be displaced by acceptance of the IEC Fieldbus are briefly introduced, with greater detail provided in Appendix A. It is noted that the IEC Fieldbus standard will be defined by a series of eight complex documents.

Chapter 5 reviews laboratory and medical automation, and discusses the serial networks used in this application area. The parallel bus has been the predominant connection method for laboratory equipment. Since short distances are usually involved this has not caused significant problems. High-energy physics research has been a very active breeding ground for this type of interconnect. Some of these bus systems have been in use for many years; for example, the CAMAC, Computer Automated Manual and Control System was developed in the mid-1960s, eventually became a BSI, IEEE and IEC standard, and is still in use for industrial and physics-related measurement and control applications. The most recent parallel bus development, the scalable coherent interface, makes specific provision for serial links; a 1000 Mbps coaxial cable link is used for room applications and optical fibres are used for longer-distance connections.

The general purpose interface bus has been successfully used since the mid-1970s. It is widely used to interconnect laboratory equipment, such as digital voltmeters and oscilloscopes, and programmable power supplies. It also has a serial extension capability (using coaxial cables and optical fibres).

In recent times electronic equipment manufacturers have tended to use the backplane interface defined by the VME standard. The VXI extension to the VME standard allows full exploitation of the instrument-on-a card concept. Systems based on this parallel bus are often used as a link between serial fieldbus instrumentation and higher-level functions which commonly involve a connection to a LAN.

1 INTRODUCTION

Advanced technology has historically had a strong influence on health care. Medical technology and the automation of medical devices and procedures have made significant advances during the past decade. Surprisingly, the ability to interface these medical devices to patient care computer systems remains hampered by the lack of an accepted interface standard. The development of a serial network standard for medical instrumentation is discussed in Chapter 5.

Considerable attention has been given to the problems associated with establishing a communication network to interconnect the large-scale health service facilities. Solutions to the problem of establishing what are in effect standard field device networks within the large central facility (e.g. the hospital) or in the smaller decentralized units such as GP surgeries, ambulance units and for patients in the home environment are less well developed.

The medical information bus (MIB) has been in development by the IEEE since 1984. It has received provisional approval as a standard by the IEEE (IEEE P 1073). It is in fact a set of three standards documents. The MIB is effectively an information pipeline that connects medical instrumentation to a host computer. It comprises two networks, one of which is an interbed multi-drop network and the other an intrabed star topology network. The MIB standard is based on the OSI communication model and it makes use of three twisted pairs: one for data, one for a synchronizing clock and one for supplying power to remote devices. An extensive user language has been defined for this bus and it has been designed specifically for medically qualified users.

Chapter 6 introduces domestic and commercial buildings as significant application areas for serial data highways and distributed intelligent sensors and actuators. This chapter will discuss the factors affecting the adoption of serial networked instrumentation.

The domestic application area is characterized by a strong technological push and a housing market that, in general, is probably not yet ready to accept high levels of home automation. A recent report (Phillips, 1994) suggests that France is actively embracing the new technology. In the USA efforts to establish a network standard started in the early 1980s but take-up appears to be slow. The small number of new houses built per year in the UK and the poor state of the housing stock will inevitably lead to a small initial market for home automation products. In Japan electronic components and new buildings are being designed with home automation in mind. If serial networks and home automation become widely accepted a systems view of the house will develop which will be beneficial to both the owner and the house builder.

Demand side management of energy consumption will motivate a limited use of serial network techniques in the home. In addition, user control of energy consumption will be facilitated by zoning areas to operate at different temperatures and with different timing cycles, arranging for the system to respond to external environmental conditions and to respond to the level of occupancy of the house. These will be the initial strong motivations to incorporate serial networked home automation into new building design and building refurbishment.

The full range of network media have been proposed and adopted for use in home automation applications, including: electricity power supply cables, twisted pairs, coaxial cables and radio links. European, American and Japanese serial network standards have been developed. All of these standards are based on a reduced stack OSI communication model. The application layer and user interface are areas of concern. To allow the construction of multivendor systems, a very large number of manufacturers will be required to agree on the format of the application layer. Since the majority of non-technical users of domestic electronic equipment find great difficulty in using this equipment it is clear that the user interface requires careful design.

The section on commercial buildings in Chapter 6 concentrates on the office working environment. Systematic premise wiring systems are available but nevertheless building wiring remains a difficult chore and cabling occupies a considerable fraction of the available space in modern buildings. Clearly serial communication networks will simplify these cabling problems and facilitate the use of information technology and automation techniques in commercial buildings.

Many serial networks have been developed for commercial building applications. Most of these are proprietary but in the future the IEC Fieldbus will no doubt find application in buildings. An important development is the use of microwave microcellular in-building communication networks. This is another area where semiconductor technology is enabling the low-cost construction of complex electronic systems. As for the domestic building the user interface must be designed with care. It is interesting to note that architects are adopting a systems engineering view of modern buildings where automation is readily accepted and serial networked communication technology is recognized as an important enabling tool.

Chapter 7 introduces measurement and control instrumentation used in transport systems. The specific areas discussed are defence systems, civil aviation and vehicle technology (including the automated road systems in which these vehicles will be used). Avionic systems have motivated the development of several bus systems. In civil aviation the ARINC serial network standard has been developed to the stage where the ARINC 629 standard now defines a complex communication system, based on OSI principles, which has been designed to cope equally well with periodic and aperiodic messages. An interesting feature of this bus standard is the specification of a current driven twisted pair with clip-on couplers. Integrated protocol circuits are available to support the implementation of this standard bus.

1 INTRODUCTION

The MIL-STD-1553 bus, discussed in Chapter 7, has been used in military avionic and other systems since the mid-1970s. It is widely supported by users and manufacturers, has been widely adopted for high performance applications in the defence industry and is noted for its reliability. MIL-STD-1553B has been adopted for use in applications ranging well beyond its original designed application area. Bus systems conforming to 1553B can be found in aircraft, missile systems, in naval systems, in tanks and in a variety of NASA space systems. This bus has a uniquely defined master device that uses a command/response protocol to communicate with other devices. This provides a good level of security but it leads to a configuration rigidity that many industrial designers of serial bus systems do not want. An early fibre optic version of 1553B was never completed, but the lower weight and natural noise immunity to electromagnetic radiation of fibre optic cable has ensured the continued development of a fibre optic version of the 1553B electrical bus. The fibre optic version of 1553B was called MIL-STD-1773 and it was released in 1987. The 1553B system is a basic system compared with the complexities of the OSI communication model and, for example, compared with the IEC Fieldbus.

The complexity of in-vehicle instrumentation has increased dramatically over the past two decades. To reduce wiring complexity and weight it is essential to make full use of serial data highway technology. Considering the size of the worldwide automotive industry it is not surprising that several serial bus systems have been developed specifically for in-vehicle use. Of the three networks discussed in Chapter 7 the controller area network (CAN) offers the best performance in error recovery and noise immunity, and its high speed capability and real-time control of information sharing between electronic control units has led to its wide acceptance. Integrated CAN protocol circuits are available from several manufacturers. It is interesting to note that the advantages of CAN have been recognized by designers working in other application areas and it has been adopted for use in industrial measurement and control applications.

Chapter 7 includes a discussion of intelligent vehicle highway systems. These systems have been developed in response to an overall system engineering approach to vehicle congestion problems, which is leading to the concept of smart cars on smart highways. The smart highway will make use of serial bus systems, and careful design will be required to ensure the successful integration of vehicle serial signals with the external signals picked up from the road system. This is an area of application for serial networks where rapid advances can be expected.

An important reason for adopting serial links to interconnect field devices has always been the need to reduce wiring complexity. This need to reduce wiring complexity applies equally well to racks of electronic equipment, printed circuit boards and integrated circuits. Field devices will usually be required to be physically small to fit into confined spaces, and in addition reductions in package size lead to reductions in manufacturing costs. Eliminating parallel bus connections from a circuit board can significantly reduce the area of the board and the overall size of the package. In many cases adequate system performance can be obtained by using serial techniques. Chapter 8 discusses the use of serial links in electronic systems, and in particular introduces the transputer as a device designed specifically for fast serial operation, and the inter-integrated circuit I^2C bus used in many consumer products.

The I^2C bus was designed by Philips for use in telecommunication, industrial electronics and consumer electronics applications. It has two bus lines: a serial data line and a serial clock line. Eight bit bidirectional data transfers can be made at up to 100 Kb/s. It is a true multimaster bus with collision detection and arbitration to prevent data corruption if two or more masters simultaneously initiate data transfer. Maximum bus capacitance is limited to 400 pF and this effectively limits the number of integrated circuits that can be connected to the bus. Each integrated circuit connected to the bus has a unique address and a simple master/slave protocol is used in which masters can operate as master-transmitters or as master-receivers. A wide range of integrated circuits are available for connection to this bus, including microcontrollers, microprocessors, memory devices, parallel-to-serial and serial-to-parallel connectors, and devices specifically oriented towards video, radio, audio and telecommunication applications. Clearly this bus has many of the characteristics of the serial field networks which have been discussed in the earlier chapters.

Chapter 9 presents a summary of the main themes of this book and emphasizes the importance of serial connection methods in the future development of automation technology. This chapter has been given the title The Connected Future because of what appears to be the unstoppable trend towards total serial interconnection of human and machine functions from world-scale telecommunication networks down to board level electronic components or even down to the circuits found in ULSICs (ultra large scale integrated circuits). In many cases machine–machine interconnections will be established and the human operator will not be involved. From a distance the world will appear to be a very large man-made brain.

A standard approach to these networks is essential. The large number of standard networks for connecting field devices discussed in this book demonstrate that the normal expectation must be for each significant market area to define its own standard. At the lower layers of the OSI model it is possible realistically to imagine an integrated circuit implementing all of the popular media access methods to be selectable by the higher layers of the model. This would reduce the duplication of hardware development but leave the more difficult problem of obtaining agreement at the application layer. Standards writing

1 INTRODUCTION

can be expected to become increasingly important in the development of serial networks for measurement and control applications.

As the use of serial networked instrumentation increases, the position of the system integrator will become increasingly important. The system integrator could also be the supplier of network devices. It is likely that the system integrator could merely be the organization that writes the software package that controls the network. In this case it is highly desirable that network devices should conform to one standard. Since the system integrator designs the user interface it is clear that he will have a major impact on the success of the network and the devices connected to the network.

The continual development of the serial network and other microelectronic technologies, along with the increasing capability of software systems is motivating the advances of manufacturing automation. Automatic guided vehicles, automatic identification and location of parts in the process of manufacture, and robot systems have all reached high levels of functionality and reliability. They all require serial communications techniques for internal and external purposes. For example, the wiring cost of implementing sensor fusion techniques in robot systems will be reduced by using serial techniques and external serial connection to higher levels of the automation will clearly be required. The system integrator will play an important part in creating flexible manufacturing systems where the rapid configuration of hardware, all under software control (i.e. the concept of soft hardware), will be used to provide a high-quality product response to changing market conditions. Such a high level of automation will lead to lower levels of employment and it is appropriate to take a broad systems view of work in our society and consider methods to maintain a position for the human operator.

Part of the work that led up to this book was supported by the UK Science and Engineering Research Council (SERC). Their support is gratefully acknowledged and particular thanks must go to the anonymous Committee that approved my application for the award of a SERC Senior Research Fellowship. I am grateful for the support and friendship of colleagues and students in the Department of Electrical Engineering, University of Edinburgh. Professor Maurice Beck in the Department of Electrical Engineering, University of Manchester Institute of Science and Technology (UMIST) has for many years provided support and inspiration for my instrumentation-related work. I have received considerable support from industrial colleagues working in the measurement and control industry. In particular, David Kent, formerly of ABB Kent Ltd., who provided friendship, support and free access to his industrial database, and allowed use of material from his 1992 Presidential address to the Institute of Measurement and Control. Special thanks are also due to the Echelon Corporation, Standard Microsystems Corporation and Phillips Semiconductors for allowing the use of material presented in Appendix C.

2
Communication Media and Codes

2.1 INTRODUCTION

A communications link (or channel) is required to enable a field-located device to exchange information with other devices and with a central control and (or) supervisory system. The channel may be a twisted pair, coaxial cable, optical or fibre optic link, or a part of the radio frequency spectrum. In some cases it may be necessary to supply power with information over the same cable; for example, a constant current with a superimposed digitally coded message and (or) data. Good design and construction practice can minimize the pick-up of interfering noise signals but it cannot eliminate it completely. A variety of digital coding and error correction procedures are available to reduce the impact on system performance of this residual noise level. The channel can be constructed to offer three modes of operation: simplex, half-duplex and full-duplex. Simplex channels offer transmission in one direction only, and half-duplex channels offer transmission in both directions but not simultaneously. Full-duplex channels provide simultaneous transmission in both directions by using dual links or a single link using a frequency division multiplexing technique. This section discusses these topics and closes with a discussion of relevant EIA (RS) channel standards.

2.2 THE TWISTED PAIR AND COAXIAL CABLE

The twisted pair and the coaxial cable are widely used to reduce noise picked up by electrical connections linking field devices. Twisted pairs can be used without difficulty up to 100 kHz, and some instrumentation applications have achieved operating frequencies as high as 10 MHz. (interference pick-up reduces as the twists per metre increases up to about 30). Coaxial cable has lower losses and can typically be used up to 100 MHz; its inherently shielded structure provides a good level of protection from capacitive pick-up.

An unshielded twisted pair provides little protection against capacitance pick-up, but it provides a good level of protection against magnetic pick-up. If cables are shielded then a braid is used rather than a solid conductor since this provides flexibility and a relatively long life. If its performance is acceptable the unshielded twisted pair provides the lowest cost and weight interconnect solutions.

Ground loops formed by earthing both ends of a link are a major source of magnetically coupled noise signals. Using a shield as an earth return and then earthing only one end of the shield provides the best results. The shielded twisted pair and the coaxial cable achieve a similar reduction of pick-up when this earthing procedure is followed. The benefit of shielding the twisted pair is quite low (about 30%) and it should be noted that the increased capacitance introduced by the shield will impair its high-frequency performance. A consequence of adding shielding is that it significantly increases the weight of the cable and in some applications, offshore oil rigs, for example, this can be a serious disadvantage.

The simple twisted pair cable is in fact a complicated physical system that requires a partial differential equation approach to obtain a detailed distributed parameter equivalent circuit. It is a transmission line and as such requires thought to be given to the use of terminating resistors to prevent reflections from line discontinuities generating unacceptable ringing conditions on every fast edge of the waveform propagating along the wire. At low frequencies a simple RC lumped equivalent circuit can be used, where R is the resistance of both conductors and C represents the total capacitance between the cables. Typically an unscreened twisted pair will have a capacitance of 50 pF/m, which increases to 150 pF/m when it is screened. Screened multicore cables will have a capacitance of the order of 200 pF/m. Cable resistance is obtained from wire tables.

The low-frequency lumped circuit becomes inaccurate when the rise time and fall times of a signal become small enough to be of the same order of magnitude as the cable propagation delay. It is commonly found that transmission line theory should be used when the 10% to 90% rise (fall) time of the signal is less than three times the propagation delay. Transmission line theory shows that when a finite line is terminated by its characteristic impedance (Z_o), it behaves like an infinite line with no reflections to corrupt signal shapes.

For twisted pairs reactive terms can be ignored, and the characteristic impedance is given by:

$$Z_o = R_o = \left(\frac{L}{C}\right)^{1/2}$$

where L is the distributed cable inductance per unit length and C is the

2.2 THE TWISTED PAIR AND COAXIAL CABLE

distributed cable capacitance per unit lengths. R_o is typically 70 Ω and propagation delay is given by:

$$t_d = l(LC)^{1/2}$$

where l is the length of the cable.

It is commonly observed that transmission line effects become significant when the length of the line is one-tenth of the wavelengths of the highest-frequency component in the transmitted signal and therefore these effects begin when the delay down the line is about three-tenths of the signal rise time. It should be noted that the fall time is often less than (e.g. half) the rise time, and it provides a more restrictive design constraint. The propagation velocity of a signal along a twisted pair line is typically 60% of the velocity of light; hence a propagation delay of 3.4 ns is equivalent to a length of 60 cm. For this length the critical rise time will be about 12 ns.

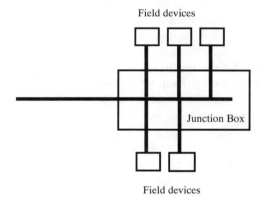

Figure 2.1
Trunk and spur, junction box electrical fieldbus topology

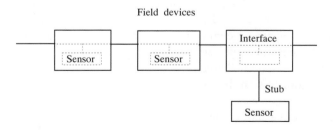

Figure 2.2
Part of a loop of serial connected field devices. Each connection is a serial, uni-directional, link

Practical systems will use a trunk and spur or a loop and spur topology, as shown in Figure 2.1 and Figure 2.2. For industrial-grade cables satisfactory operation over a trunk length of 100 m can be achieved at 1 Mb/s and spur lengths less than 6 m can be used without a terminating resistor. With premium-grade cables the trunk length can be 400 m at 1 Mb/s. Much longer trunk connections can be used as the bit rate is reduced; for example, at 62.5 kb/s the trunk length can be 1900 m and 50 m spurs can be used without terminating resistors.

2.3 FIBRE OPTIC NETWORKS

Early optical fibres exhibited an attenuation of the order of 1000 dB/km and this high value was due to the high level of impurities in the glass. After 25 years of continuous development, fibres with attenuations as low as 0.2 dB/km are now commercially available. Monomode fibres can offer bandwidth distance products values of 2000 Gbit km and they are clearly ideal for long-haul communication links. However they are not suitable for field-instrumentation or short-distance applications because, for example, it is difficult to launch optical signals into the small diameter (usually less than 10 μm) of monomode fibres. In multimode fibres the core diameter exceeds approximately ten wavelengths. In this case the number of transmission modes is considerable and it approaches a continuous distribution. Multimode step and graded index fibres, and their connecting components, have reduced in price to the point where applications in instrumentation systems can be attractive.

Fibre optic links offer several advantages. They are immune to electromagnetic interference. They do not emit signals, so high security can be guaranteed and cross-talk between fibres is not a problem. Low power (mW or less) signalling is easily achieved and fibre optic links (when correctly installed) can offer high reliability. In addition a fibre link can offer low weight and small size, especially when compared with electrical connection methods. Of course fibre optic technology is not without its own difficulties. For example special techniques are required to terminate links, coupling losses are usually higher than with electrical connections and careful path routing is required to ensure that a fibre is not subjected to sharp bends. Surprisingly, after 25 years of development, fibre optic technology is still often regarded as a new technology for instrumentation system builders and maintenance engineers to master.

An optical fibre consists of a highly refractive core surrounded by a cladding of a lower refractive index material. A beam of light entering the core of the fibre is transmitted by total internal reflection in the case of a step index fibre, or multiple refraction in the case of a graded index fibre. In a step index fibre light can travel down the fibre at different angles with respect to the axis of the

2.3 FIBRE OPTIC NETWORKS

fibre core and for each angle a different path length is obtained. Consequently light will arrive at a receiver at different times, causing pulses to broaden, or equivalently, limit the bandwidth of the step index fibre. In a graded index fibre light entering at different angles is gradually refracted until total internal reflection takes place. Light then continues to be refracted and reflected in a symmetrical manner which leads to a uniform path length and a bandwidth greater than can be achieved with the step index fibre.

Polymer and glass (silica) fibres are available for point-to-point and serial network applications. Polymer fibres can transmit up to 200 m at up to 100 kb/s (or 100 m at 10 Mb/s). In general they are used in short-distance, low-bandwidth and low-cost applications. Silicon fibres are used for high-speed (hundreds of Mb/s) and long-distance applications. They are available in many forms, including: single-mode fibres for very high speed and distance telecommunication application; multimode graded index, with cores up to 100 μm in diameter, for medium distance and bandwidth applications; and multimode step index fibres, with cores of up to 200 μm diameter, for radiation-resistant optical communication networks and low cost and bandwidth applications. The optical characteristics of polymer and glass fibres are shown in Table 2.1. In general larger-core fibres have higher attenuation than similar-material small-core fibres. Graded index fibres have a wider bandwidth than step index fibres (with the largest core fibre having the smaller bandwidth). Typically, a 50 μm fibre has a bandwidth–distance product of 400 MHz km, while a 200 μm fibre has a bandwidth distance product of 20 MHz km. Note that the cladding of the fibre largely governs its cost, which can vary over a 20 to 1 range from armoured cladding (suitable for direct burying) to a fibre with minimal sheathing for indoor use.

Table 2.1
Fibre optic transmission windows

Fibre	Wavelengths Window	Comments
Polymer	565 nm (green visible light)	GaAlAs double heterojunction, ultrabright LEDs, short distances < 250 m. Polymer fibres not suitable for IR transmission, except over very short distances
	660 nm (red visible light)	
Glass	800–900 nm (infra-red)	Up to 10 km with data rates up to 200 Mb/s
	1300 nm (infra-red)	Low attenuation, single-mode 0.3–0.8 dB/km, multimode 0.5–1 dB/km

The power which is launched into a fibre from a given source is approximately proportional to the square of the diameter of the fibre. For example, if 1 μW can be injected into a 50 μm fibre, a 200 μm fibre will accept 16 μW under similar circumstances.

A wide range of fibres is available. Single-mode fibres with diameters of about 9 μm and working at 1300 nm are used in long-haul high-bandwidth, applications because their attenuation is extremely low (< 0.2 db/Km). Because of the small diameter it is difficult to launch significant power into the fibre and to align the fibre with the source. Multimode fibres with diameters of 50 or 62.5 μm have low attenuation (3 db/μm or less) and it is easier to launch

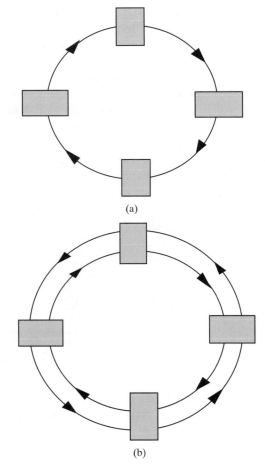

Figure 2.3
Basic fibre optic ring topology

2.3 FIBRE OPTIC NETWORKS

significant power into the fibre. Multimode fibres with diameter of 100, 200, 400, 600 μm, etc., have higher attenuation (of the order of 6–15 dB/km) and lower bandwidth (about 20 MHz km). However it is relatively easy to launch high power into these fibres. Fibre optic technology continues to develop, so further improvements in performance and reductions in cost can be expected.

Several topologies have been investigated for fibre-optically linked instrumentation. The simplest and probably the most widely used in practice is the ring of repeaters shown in Figure 2.3. Relatively simple photodiode (PD) receivers and light-emitting diode (LED) transmitters are required to implement this type of connection. PD/LED combinations can be used to form half-duplex links with fibre optic Y splitters and twin fibre duplex links. Note that the simple

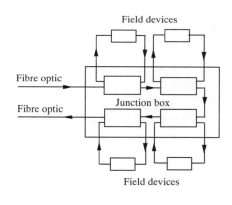

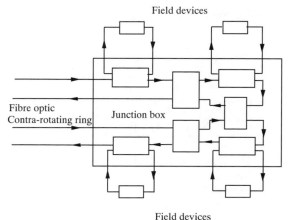

Figure 2.4
Fibre optic ring topologies using a junction box and two-input/two-output repeaters

ring is very susceptible to the failure of any element in the ring. One partial solution to this problem is to use two contra-rotating rings (Figure 2.3) but care is still required to minimize the effect of common mode failures.

Many industrial users will not want to loop a fibre around machinery to form a ring connection, and will in most cases require a junction box approach to be followed, similar to that used with the copper (wire) connection methods. Also, the insertion (or removal) of devices from an active ring is inconvenient. Figure 2.4(*a*) shows a fibre optic ring entering a junction box which uses two-input/two-output repeaters to connect field devices to the ring. In this

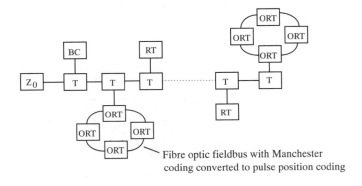

Fibre optic fieldbus with Manchester coding converted to pulse position coding

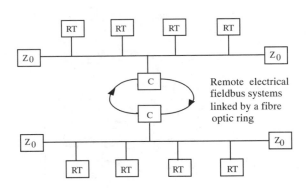

Remote electrical fieldbus systems linked by a fibre optic ring

C = optical / electrical converter
BC = Bus controller
RT = Remote Terminal
ORT = optical remote terminal
T = Transformer coupler

Figure 2.5
A fibre optic fieldbus combined with an electrical, twisted pair, Multidrop bus

2.3 FIBRE OPTIC NETWORKS

case devices can be added or removed from the subrings without affecting the overall operation of the ring. Of course the controlling software system must be designed to react automatically to the absence or presence of devices. Figure 2.4(b) shows a contra-rotating ring system coupled using two-input/two-output repeaters in a junction box configuration, and in this case failure of a repeater can be overcome by using software to instruct the immediately adjacent functioning repeaters to modify their internal connections to form two non-overlapping rings. Hence only a small part of the network is affected by the failure.

As noted above, glass fibre optic links can transmit wide bandwidth signals over relatively long distances but polymer fibres are only suitable for short distance links. Hence a glass fibre will be required to connect to junction boxes over long distances, while polymer fibres can be used for short-distance connections in the immediate vicinity of the junction box. An alternative arrangement would be to combine an electrical serial bus with a fibre optic bus as shown in Figure 2.5. In this case two electrical bus systems could be linked by a fibre

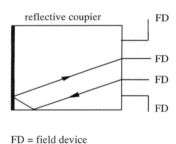

FD = field device

Reflective star

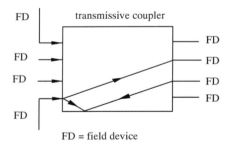

FD = field device

Transmissive star

Figure 2.6
Optical passive reflective and transmissive stars

optic ring, or a ring extension could be added to a multi-drop electrical bus. These ring extensions can be designed to operate with very low power and they are therefore attractive for operation in hazardous locations when intrinsic safety is required.

Fibre optic connection topologies based on optical stars and trees have received considerable attention. Y splitters can be used to enable half-duplex links to be formed. Passive reflective and transmissive splitters have been developed. Schematic diagrams of these devices are shown in Figure 2.6 and their use in a multistar topology is illustrated in Figure 2.7. Cruickshank and Kennett (1989) have reported the use of a local star topology implemented in a transmission format to link helicopter avionic systems. Progress has also been made with reflective star couplers. For example Graves (1989) has reported the use of fused fibre technology to produce a 32-port reflective star with a maximum loss of 19 dB. Clearly this is an area where technological development can be expected to improve the capability and attractiveness of fibre optic systems.

The simplest systems are those that use point-to-point communication, requiring components that are readily available. Each link consists of a transmitter, a receiver and an optical fibre which may have in-line connectors at bulk heads or when the fibre has to be extended. For field instrumentation applications data rates are likely to be of the order 1 Mb/s. For links of the order

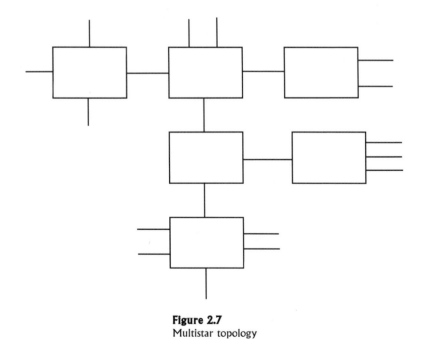

Figure 2.7
Multistar topology

of kilometres, 50 or 62.5 μm fibre is commonly used by industry, because it has about half the cost of 100 or 200 μm fibre and it is easy to launch sufficient power into this diameter. However, if low power consumption transmitters and/or simple low-cost receivers are a requirement, then high-diameter fibres will be of particular value.

In the interests of low cost and circuit simplicity, the transmitter will be an LED with a current-limiting resistor driven by the output of a logic gate. If transmitter power has to be kept low, the signalling technique could use narrow pulses of the order of 100 ns in duration. Under these circumstances the bandwidth, and probably the cost, of the receiver will increase. Recent advances in the manufacture of GaAlAs light-emitting diodes give greater power output at lower drive currents. For example 70 μW at 880 μm can typically be launched into a 50 μm fibre by a commonly available diode when driven by 100 mA.

PIN diodes are the lowest-cost choice for a receiver. No recent significant technical advances have occurred in sensitivity, which still remains at about 0.5 to 0.6 A per watt for 850 nm devices. Capacitance (which determines response times) and cost appear to be the only parameters available for further reduction. Using a PIN diode it is possible to design a receiver which has a sensitivity of the order of 100 nW. However it must be appreciated that a receiver can be swamped by too much input power. This can cause severe saturation of the preamplifier and such extreme signal distortion that data is corrupted. In a large-scale instrumentation system, link lengths will vary over a wide range, and the resulting variation of fibre attenuation may cause problems with short links if steps are not taken to limit transmitter power and/or receiver sensitivity in these cases. The simplest electronic solution to this problem is to control the transmitter power by altering the current-limiting resistor which feeds the LED.

It may suffice to have a stock of two transmitter types, i.e. high power and low power.

It is interesting to note that most LEDs also act as photodiodes when reverse-biassed. Simple tests show that a LED used in this way has a sensitivity of about 5% of a normal photodiode. Even with this 13 dB loss of sensitivity, viable single-fibre half-duplex links could be served by LEDs acting as transmitter/receivers, provided that transmitter current is kept high and the link length is not extreme.

2.4 THE RADIO LINK

To interconnect field instrumentation with a cable is an expensive man-power-intensive operation. In some circumstances cabling is technically difficult, as is the case when the device to be connected is moving or rotating. Alternatively instrumentation could be widely dispersed and then use of technical staff to collect data using hand-held equipment can sometimes be justified, but the reliability of this method is low, especially over long periods of time. Hence there is a need for a low-cost method to collect signals from remote or mobile instrumentation without the need for interconnecting cables. The application areas for cableless signal transmission include:

- mobile or moving equipment
- temporary instrumentation and rapid prototyping
- pipeline instrumentation
- environmental monitoring
- instrumentation in hazardous environments.

The need for more on-line data capture continues to increase. A large part of this increase is associated with instrumentation in situations where direct links are not feasible and, commonly, a direct line of sight cannot be easily established.

In applications where line of sight can be established, then infra-red and acoustic techniques are often used to link instrumentation systems over distances less than about 50 m. Over much longer distances (greater than about 3 km) the traditional radio (microwave) line of sight link is used. Factory automation and process control applications often involve non-line-of-sight links, and require very low power consumption by the remote devices to allow power to be supplied by battery or solar sources. Figure 2.8 summarizes the factors that lead to the choice of the VHF/UHF bands for the radio fieldbus.

Radio telemetry has a long and successful history of measurement and control applications. Interest in radio telemetry for measurement and control has increased since the success of the digital techniques used in pagers and cellular radio telephones. Neve (1990) defines three classes of radio telemetry to connect field devices:

- low data rate (e.g. 2400 b/s) for low power operation with battery or solar power supplies

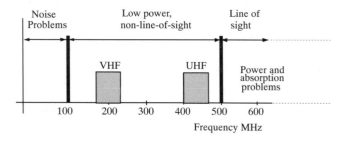

Figure 2.8
The VHF/UHF telemetry bands

2.4 THE RADIO LINK

- high data rate (e.g. 20 kb/s) for applications where power consumption is not critical, e.g. pipeline automation
- high operating frequency (greater than 800 MHz) for use in buildings with steel reinforcing.

A related activity that is developing largely independently of the radio fieldbus is the wireless LAN (WLAN) to connect mobile computing platforms (e.g. laptop and desktop equipment). Standards are being developed (e.g. IEEE 802.11) that define communication between these computing devices using multiple client–server and peer-to-peer network operating methods. National Semiconductor has proposed a unified media access control protocol which combines the major elements of these systems. The trend to use the ever-increasing capability of digital silicon technology allows multiple solutions to be implemented, but it also allows competing groups to escape from the need to work for a single consensus solution. It remains to be seen if higher-level system software can be produced that will enable standard interfaces to be established with multiple solution silicon protocol circuits.

Notwithstanding these difficulties telemetry manufacturers have established nationwide networks for the transmission of digital data. For example PAKNET, a joint venture between Racal Telecom and Mercury Communication, is a cellular network designed specifically for digital data transmission. This network consists of a matrix of base stations which are connected to electronic data exchanges by high-speed digital links. Data is sent over this network in short (27 ms) segments, called packets, using the industry X25 packet switching standard (Tabagi and Kleinrock, 1976; Folts, 1980). Base stations communicate with other stations over leased lines and they communicate over longer distances with VHF radio links. Figure 2.9 shows a diagram of a packet radio data network (PAKNET). At the remote site data is collected and, using a standard protocol, it is fed into network terminating units (NTUs) which then transmit the encoded information to the nearest base station. At each base station there is a direct link into one of the packet switch exchanges which form the core of the X25 network. Information is received by using another NTU, or by a dedicated line, or via the telephone network. Host computers can be connected to the network either directly via digital lines or through other X25 data networks.

PAKNET effectively implements the first three layers of the OSI communication model (Baddoo and Martin, 1991). A radio frequency allocation of 159–164 MHz has been granted and the spatial reuse cellular technique has been adopted with a seven-cell reuse pattern. Channel bandwidth is 12.5 kHz and the frequency shift keying (FSK) modulation scheme is used. Link data rates of up to 8 kb/s can be achieved.

The remainder of this section will discuss the main features of the radio fieldbus option proposed for the IEC fieldbus standard.

2 COMMUNICATION MEDIA AND CODES

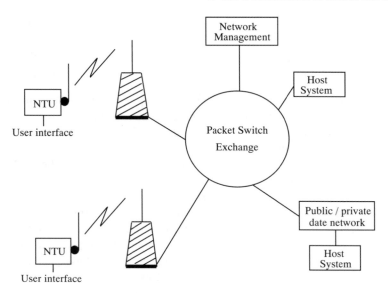

Figure 2.9
Packet radio data network

The draft low-speed radio-medium physical layer is mainly aimed at the requirements of the process industries. It has been designed to enable implementations of the standard to be functionally indistinguishable from a directly connected fieldbus, except for a slow response. Non-line-of-sight operation with small integral antennae will be possible, and with sufficiently low power levels to enable conformance to intrinsic safety (IS) and flameproof (ExD) standards. Operating distances will be up to 4000 m, and for low-power operation 40 m and 400 m ranges can be specified. Reduced power consumption will also be achieved by using small operating duty cycles. Channel bandwidth will be 12.5 kHz and the transmitted bit rate will be 4800 b/s. It should be possible to send at least ten measurements per second over a single radio fieldbus.

National regulatory agencies (e.g. the Radio Communication Division of the DTI in the UK, and the Federal Communication Commission (FCC) of the USA) allocate segments of the frequency spectrum for use in radio applications. Since process control applications require non-line-of-sight links with low-power operation the radio fieldbus must use the low-power telemetry frequencies.

As shown in Table 2.2 these frequencies are different in most countries. By using the minimum channel bandwidth of 12.5 kHz with a modulation technique (Gaussian minimum shift keying, a form of FSK) that reduces out-of-band spurious radiation the regulatory restrictions effectively set the range of

2.4 THE RADIO LINK

Table 2.2
Examples of low-power telemetry bands

Country	Frequency (MHz)	Maximum power (mW)
UK	173 (VHF)	1
	458 (UHF)	10 (Lic.)
Netherlands	153 (VHF)	500
	450 (UHF)	500
Norway	142 (VHF)	500
	442 (UHF)	500
France	152 (VHF)	5
	450 (UHF)	100 (Lic.)
Denmark	225 (VHF)	10
	448 (UHF)	100

the communication link for a given receiver sensitivity. Some of the bands will not require a license. The draft standard allows for licensed and non-licensed applications by enhancing the error-handling procedures in the physical layer. If an interfering user remains a problem the result will be more retries and a lower channel throughput.

By specifying that the transmitter power should be set to the maximum allowed by the appropriate regulatory body, receiver performance becomes the design parameter that can be adjusted to provide different levels of overall system performance. From Table 2.2 it can be seen that receiver specifications will depend on the country in which the fieldbus will be used. Performance will of course depend on the characteristics of the communication channel. For applications in the process control environment a Rayleigh fading characteristic is commonly experienced. In short-range applications a low-cost insensitive

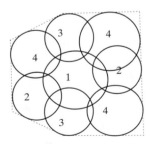

Figure 2.10
Example of a microcellular topology for a site using radio fieldbus techniques

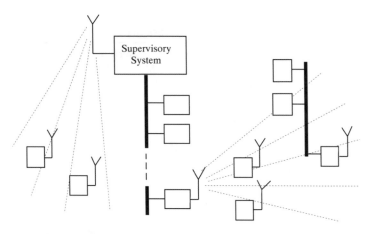

Figure 2.11
Schematic diagram showing the application of the radio fieldbus

receiver can be used without any error rate penalties. Since it is undesirable for the longer-distance applications to lead to over-specified receiver designs for shorter-range applications, three conformance classes have been specified in the draft standard, namely 0–40 m, 0–400 m and 0–4000 m.

The allocated frequency bands for low-power telemetry (VHF and UHF) allow for at least four channels to be implemented, and transmitters can be arranged in a microcellular topology as shown in Figure 2.10. Individual radio fieldbus systems can be implemented; for example, as shown in Figure 2.11. A specific radio fieldbus will be defined to operate in a geographical area (a cell). Networks in adjacent cells will use different radio channels.

2.5 4–20 mA ANALOGUE SIGNALLING AND INTRINSIC SAFETY

For analogue signalling, current is generally preferred since the low impedance of current-operated circuits leads to a lower level of noise interference. This lower impedance also leads to faster circuits and hence wider bandwidth. The live zero 4–20 mA signalling range has been used successfully for many years. British Standard 3586 appeared in 1970. Twisted pair cables are used to reduce the effect of electromagnetic interference, and if electrostatically induced (i.e. capacitively coupled) noise is a problem, then the twisted pair is surrounded by a screen. A circuit diagram for a two-wire live zero loop is shown in Figure 2.12. Note that the device that sets the loop current is usually called a transmitter. A wide range of sensors are available in transmitter configurations.

2.5 4-20 mA ANALOGUE SIGNALLING AND INTRINSIC SAFETY

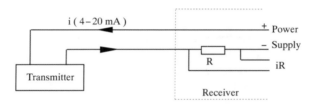

Figure 2.12
4-20 mA analogue signalling

Note that the lower 4 mA is the effective power supply for the transmitter, and circuits are required to convert this current into a voltage supply. A major difficulty of transmitter design is the achievement of low-power and low-voltage operation of sensor circuits.

Intrinsic safety (BS5501, part 7) is an important feature of the design of instrumentation for use in hazardous process control applications. Design for intrinsic safety is based on the fact that a certain minimum energy is needed to ignite an explosive mixture of gases with air. An intrinsically safe field device is designed to contain only a small stored energy, usually in capacitors and inductors, but note that energy storage in field wiring can be significant. Also, voltage and current is supplied to a device via an energy-limiting barrier, so that even under fault conditions the power is limited by a safely specified maximum voltage and maximum current. The limits are chosen to give an acceptable safety margin below the energy required to cause an explosion. Different gases form explosive mixtures with air which have different ignition energies. Groups of gases are defined to relate gases with similar explosion energy limits. Intrinsically safe devices are categorized according to the severest gas groups in which they may be used. The European Committee for Electrotechnical Standardization (CENELEC) defines the following groups (in increasing hazard order):

Group IIA: ammonia, propane
Group IIB: ethylene
Group IIC: hydrogen, acetylene

A device designed for use in hazardous environments is issued a Certificate of Conformance by national approval bodies (e.g. BASEEFA in the UK).

The use of a barrier to supply a 4-20 mA loop is shown in Figure 2.13. A variety of barrier circuits have been developed. The basic circuit, shown in Figure 2.14, uses shunt zener diodes to limit voltage and current-limiting resistors with a fuse which blows in the event of over-voltage being applied by the power supply. This basic circuit drops a significant voltage across its series resistors and this, in combination with the voltage dropped across

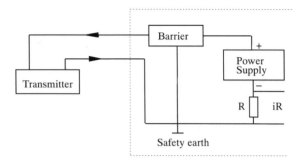

Figure 2.13
Use of a barrier to prevent explosions in hazardous areas

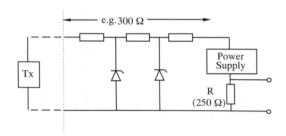

Figure 2.14
Zener (shunt diode) barrier

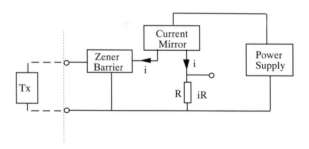

Figure 2.15
Zener barrier with a repeater circuit used to increase the voltage across the transmitter

the connecting cables, may not leave sufficient volts to supply the transmitter circuits. Figure 2.15 shows a solution to this problem which uses the current mirror circuit technique (sometimes called the repeater technique) to separate the zener barrier from the load resistor.

2.6 DIGITAL CODES

Ground loops are a major problem commonly experienced when wiring field devices. The isolating repeater barrier provides galvanic isolation for the power supply to field devices, usually by means of a DC-to-DC convertor using a transformer with well-separated windings.

Since isolation can be guaranteed there is no need for safety ground connections.

2.6 DIGITAL CODES

The direct transmission of a bit stream (in the non-return-to-zero (NRZ) form) is not used because of the difficulty of recognizing bit divisions in sequences of 1s (or 0s) in the received signal. A return-to-zero (RZ) representation can be used, but its implementation involves more complicated circuits and requires twice the bandwidth used for the direct technique. The Manchester encoding method provides a more convenient representation of a digital bit stream. It solves the problem of recognizing bit divisions in sequences of 1s (or 0s) and provides transitions in each successive bit that enables the easy implementation of clock recovery. The Manchester method is widely used in serial network standard specifications. Figure 2.16 shows waveforms to illustrate these digital encoding methods. The frequency content of the encoded signal should have a

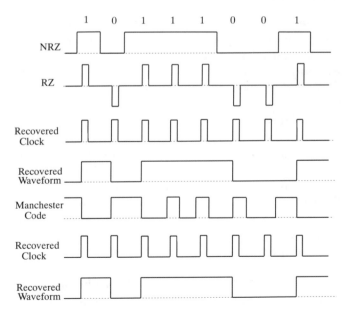

Figure 2.16
Waveforms to illustrate digital encoding and clock recovery

zero DC component, minimal high-frequency components and low bandwidth. In addition it is desirable that the cable connections should be non-polarized, i.e. it should be possible to interchange conductors without affecting data transmission. However the widely used 4–20 mA analogue signalling method is polarized and it appears to be used without difficulty, so the non-polarized connection capability may not be a practical requirement.

Like all digital signals the Manchester encoded signal will be distorted by propagation down a long twisted pair. It is not a simple matter to correct this by predistortion at the transmitter or equalization at the receiver. Predistortion is the commonly used technique. The Manchester encoded signal does not have a component at the bit rate frequency which can be used to recover the bit clock rate. Methods of clock recovery which depend upon the timing of zero crossings are commonly vulnerable to noise and timing filter. It is desirable that a standard specify a single design approach, the same modulation and encoding technique for all network topologies and media and function requirement variations. The encoding scheme should offer clock recovery, otherwise the physical layer becomes dependent on the data link layer, thereby losing conformance to the OSI layered approach to standardization.

2.7 ERROR DETECTION AND CORRECTION

Interfering noise signals, which often come in bursts rather than as single events, cause errors to be generated in the transmitted digital bit patterns. Consequently a communication channel can be expected to suffer from a group of successive error bits rather than a single bit error. Two approaches to error control have been developed, namely, forward error control and feedback error control. With forward error control redundant bits are added to the basic digital signal before transmission. The redundant information enables the receiver to detect when errors are present, and deduce the changes that should be made to the received signal to enable the correct information to be reconstructed. In contrast, with the feedback error control method the additional redundant information is used by the receiver only to detect errors. When an error is detected the receiver requests a retransmission, and if errors continue to be detected the receiver will indicate a system fault condition. The number of additional bits required to achieve reliable error correction increases rapidly as the number of message bits increases; hence feedback error control is the widely used technique.

If the allowed code words can differ by only one bit then a single bit error will not be detectable, since the erroneous word created will in fact be a valid code word from another position in the code table. To overcome this problem techniques have been developed that allow code words to be created, with the

2.7 ERROR DETECTION AND CORRECTION

property that bit errors do not generate other words from the code table. The term distance is used to represent the number of bits that can be changed before another valid code word is obtained.

The distance concept can be applied to the correction of errors. A minimum distance of three between code words must be used to implement the detection and correction of single errors. If a message is received that is not in the code table the receiver system effectively selects the code word nearest to the received message. When the minimum distance is three, either single bit error correction, or double error detection, but not both, can be implemented. If the minimum distance is increased to seven then the receiver can be designed to detect and correct three errors.

A parity check code is a widely used method to increase the minimum distance of the basic digital code. A single parity check code, which gives a minimum distance of two, involves the addition of a parity bit to each word. An old parity bit is added if it makes the overall bit count an odd number. This is the usual technique to be adopted since it precludes the reception of an all-0s word. The parity technique can be extended to apply (vertically) to a block of words as well as (horizontally) to a single word. This type of code is referred to as a block parity code and it is particularly useful to detect errors caused by a burst of noise.

The Hamming code is widely used both to detect and correct errors by means of multiple parity checks made upon specific bit positions in the transmitted word. If parity checks are performed and the total length of the transmitted word then becomes $(m+k)$ where m is the number of bits in the original word. The k parity checks are formed into a k digit word that is capable of defining a maximum of $2k$ different states; one of these states indicates that the word is correct with all parity bits at logic 1. Hence the minimum value of k that satisfies:

$$2^k - 1 \geq m + k$$

is used to construct the Hamming coded word.

The simple parity (or block parity) codes do not provide reliable protection against bursts of noise. In this case error control is usually achieved by the use of a polynomial code. These codes are also referred to as convolutional codes and the generated check bits are called a frame check sequence (FCS) or a cyclic redundancy check (CRC).

The following operations are required to implement the polynomial error detection method. Assuming that the FCS is an n-bit word, the message sequence is first shifted n places left (i.e. multiplication by 2^n). The FCS is the remainder obtained when the shifted message sequence is divided by a special digital sequence called the generator polynomial. An example of a generator polynomial is given below:

$$P(x) = x^{16} + x^{12} + x^5 + 1$$

(this indicates a 17 bit word with binary 1 in positions 17, 13, 6 and 0). A divisor which is prime (in the modulo 2 sense) is normally used. If the transmitted signal is received without error then a zero remainder is obtained when division by the generator polynomial is again implemented by the receiver. A non-zero remainder indicates an error. Error detection and correction codes have received considerable attention (see, for example Welsh (1988) and Stallings (1991)).

2.8 SPREAD SPECTRUM ENCODING

Very often radio telemetry and power line communication links need to operate reliably even when high levels of interfering noise can corrupt the transmitted signal. A spread spectrum system uses a transmitter bandwidth much wider than the data to be transmitted and this enables a high signal-to-noise ratio to be maintained without increasing transmitter power. The spread spectrum encoding method is based on the use of pseudo-random noise signals (mixed with the signal to be transmitted) and correlation detectors. Different pseudo-random codes are used with very low cross-correlation coefficients to enable several transmitters to have simultaneous access to a communication channel. Two techniques are commonly used in a spread spectrum system; namely, the frequency hopping method and the direct sequence method.

A frequency hopping transmitter switches round a defined range of channel frequencies in a sequence defined by a pseudo-random noise generator. The time it dwells on a particular frequency is very short so the transmitted signal appears to be a burst of interfering noise. Frequency shift keying is often used to modulate the digital data onto the carrier. Digital data can be generated by conventional convertor circuits or a continuous bit stream can be obtained by delta sigma modulation methods. The receiver uses identical frequency channels which are randomly selected by the pseudo-random sequence generator. A synchronization circuit in the receiver locks its pseudo-random code generator to the transmitter code generator. Hence the broadband, low-power, transmitted signal can be decoded to obtain the original digital data.

In a direct sequence spread spectrum system the pseudo-random code phase shift keys the carrier. This of course increases the bandwidth (i.e. it spreads the spectrum). Typically a code inversion method is used where logic 1 of the code selects no phase shift while logic 0 selects 180° phase shift (i.e. an inversion). This is achieved in practice by using a double-balanced mixer (DBM) to switch the carrier between 0 and 180° phase shift (this is known as bi-phase shift keying, BPSK, or phase reversal keying, PRK). Digital data is exclusively ORed with the pseudo-random spreading code (this inverts the modulating coding if data is 1). The receiver circuit uses an identical synchronized code sequence to despread the direct sequence signal. A bi-phase shift demodulator is then used to recover the original digital data. As with the frequency hopping arrangement, a complicated synchronization circuit is required for the successful operation of this method.

Hoshikuki *et al.* (1992), have reported a spread spectrum system that shows that a frequency hopping circuit can operate without a complicated synchronizing circuit. Their system features a hybrid direct sequence (DS)/frequency hopping (FH) synchronization scheme. The hopping time interval is designed to coincide with the DS spreading sequence cycle, and therefore the frequency switching time coincides with the peak generated by the DS correlator. To maintain synchronization any difference between the correlation peak timing and the FH switching timing in the receiver is sensed and compensated for by the hopping controller. High-speed synchronization is possible because the acquisition time of this system is about one hopping cycle.

Marshall and Spracklen (1990) discuss the use of direct sequence spread spectrum methods for implementing a LAN using fibre optic media. They propose the use of a star topology with an active hub which linearly adds signals to produce a composite network signal which is then broadcast to all nodes connected to it. The large processing gain and simultaneous multiple access capability of the spread spectrum technique gives the proposed network a high degree of noise insensitivity and message privacy coupled with a fixed latency irrespective of traffic density.

Spread spectrum communication methods are making an important contribution to the successful use of power lines as communication links. This will be discussed in the next section.

2.9 POWER LINE COMMUNICATION NETWORKS

Power line signalling eliminates the need for communication cables. Clearly this significantly reduces the cost and simplifies the task of retrofitting serial networked instrumentation in existing buildings. At the time of construction of a new building it is relatively easy to install communication cables and this will be the preferred approach. Nevertheless even in a new building the power line communication method provides a useful extra level of flexibility for users.

Unfortunately the power line suffers from very high levels of interfering noise, and in addition it has electrical characteristics which lead to poor transmission performance at useful data rates, especially when electrical devices are connected to the power line. Attenuation of a transmitted signal can easily reach 60 dB. The interfering noise typical arises from devices such as motors, switching power supplies, light dimmers and fluorescent light ballasts connected to the line. Since devices are constantly being switched on and off the noise and the electrical performance of the cable are continuously changing. Clearly it is unlikely that simple communication methods will be successful, and sophisticated error detection and correction methods are likely to be required.

Like the radio frequency spectrum the power line bandwidth is restricted by National Agencies. In Europe long-wave radio stations severely limit the upper frequency. Consequently the frequency range 9–95 kHz is reserved for utility applications such as remote meter reading and load shedding. The band 3–9 kHz is also restricted for use by electricity suppliers but it may also be used for signalling in consumer installations under conditions authorized by the electricity supplier. Europe has also allocated the frequency range 125–140 kHz for non-utility general-purpose applications. In addition strict limits are imposed on the shape of the signalling waveform. The bands 95–125 kHz and 140–148.5 kHz are reserved for consumer use, but for these bands an access protocol is not required.

The Federal Communications Commission (FCC) set the allowed power line signalling bandwidths in the USA, where it is AM radio operating down to 535 kHz that limits the highest allowed frequency to 450 kHz. In the USA power line noise increases significantly below about 100 kHz, so a 350 kHz band is available for practical use. The first power line signalling British Standard, BS 6839, was published in 1987. This is now superseded by BS EN 50065, 'Signalling on low-voltage electrical installations in the frequency range 3–148.5 kHz', which was published by CENELLEC in 1992.

Devices designed to communicate over the power line must be designed to be safe over a wide range of conditions. A doubly insulated transformer is used to couple carrier frequencies to the power line. In addition the interface circuit must be protected by using filters to eliminate the power line frequency component and back-to-back diodes are used to suppress transients that could damage carrier interface circuits. Since power line failures are relatively common, applications requiring total security should not use the power line as the sole communication link.

Early power line signalling specifications (including the first CEbus standard) were based on amplitude shift keying (ASK). A commercial system of this type was introduced by X-10 Inc (USA) which used a 120 kHz carrier synchronized to transmit one logic bit at every zero crossing of the power line signal (60 kHz in the USA). A logic 1 was denoted by a 1 ms burst of the 120 kHz carrier and a logic 0 was denoted by a no-signal period of 1 ms. The CEbus used the same carrier, and to prevent conflict with previously installed X-10 devices it was found necessary to introduce an extra symbol into the CEbus protocol.

2.9 POWER LINE COMMUNICATION NETWORKS

Unfortunately this also had the effect of wasting nearly 50% of the channel bandwidth. It is clear that since access to the power line is freely available, great care must be taken when introducing new products to ensure that they can co-exist on the power line bus with previously developed devices.

In domestic applications the distribution transformer is a common feature of the power network that supplies a group of houses. Any house in the group can communicate, via the power line bus, with any other house connected to the distribution transformer. Each will require an identification address that ensures privacy and prevents involuntary interference from other houses. Attention must also be given to the detailed wiring, e.g. where different phases of a three-phase system are used (with respect to neutral) or connections are across two phases of a three-phase system. Communication across phases will require additional signal coupling components to be installed. A passive bridge can be implemented for a transformer by connecting high-voltage capacitors across the transformer; alternatively connectivity across a transformer can be provided by repeaters or store and forward devices. Installing a power line communication link will not always be a simple matter of plugging into a power line socket.

Power line communication using amplitude shift keying or frequency shift keying is difficult to maintain because of the predominantly impulsive (rather than white) nature of commonly experienced interfering noise (Dostert, 1992). Practical experience has shown that data rates of a few kb/s can be achieved with an acceptable bit error rate. Unfortunately the communication link may fail when electrical equipment (e.g. an electric drill) is connected (or disconnected). A solution to these noise problems can be found by adopting the spread spectrum techniques discussed in the previous section.

Dostert (1988) describes a frequency hopping spread spectrum system for power line communication where up to 30 users can simultaneously occupy the network; each at a data rate of 300 b/s. A simple link was shown to operate for ten hours with no errors in a one-family house located in a rural environment. In a more testing laboratory environment, bit error rates up to 10^{-3} were achieved. In a more recent paper, Dostert (1992) describes the use of ASIC technology to implement a frequency hopping spread spectrum modem which will fit into the housing of a standard wall-socket with a power consumption of a few hundred milliwatts.

Saund, Comley and Hill (1990) describe a 488 b/s power line communication system that uses the direct sequence spread spectrum technique. The frequency band was selected to be 300 kHz to 1.3 MHz. This band is outside the currently allocated bands for mains signalling and will not interfere with already established power line links. A pseudo-random code, 1023 bit maximal length sequence was used with a clock rate of 500 kHz and the carrier frequency was 800 kHz.

Power line spread spectrum systems are commercially available. The Echelon Corp system is based on a modified direct sequence spread spectrum technique which uses 31 chips/bit for encoding data to be transmitted. A chip is the smallest pulse width in the encoding pulse sequence used to represent a bit. The number of chips/bit employed by the encoding scheme is a trade-off between the degree of spectrum spreading and bit rate. The 31 chips/bit encoding scheme yields bit rates of 10 kb/s for the 100–450 kHz band available in the USA and 2 kb/s for the 9–95 kHz band used in Europe.

Gershon, Propp and Propp (1991) discuss the key features of a data link layer required for reliable operation of large power line communication networks. They note that only a certain amount of contiguous information can be sent before it is almost certain that transmission will be corrupted. This suggests a requirement for transmission of short frames over the power line. To ensure further the integrity of any frame of data it is necessary to use both error-detecting and error-correcting codes; forward error correcting to minimize the number of retransmissions and error detection to know if there is a need for a retransmission on a frame basis. Each frame should be acknowledged by the receiver before the transmission proceeds to the next frame. To implement this low-level link protocol, the higher-level packet is broken up into short frames. Power line conditions can change of the order of every few milliseconds, hence the receiver must be able to adapt to these changing conditions. Using a low-level link protocol built upon short frames, the receiver can adapt on a frame basis and, because acknowledgements are required, no information is lost. This requirement for short power line frames can be met by a physical layer spread spectrum technology that provides for rapid synchronization.

2.10 THE EIA RS STANDARDS

The RS series of standards define the electrical characteristics of a serial interface. They do not specify a modulation method or message formats. The RS-422(423) standards appeared in 1975 and defined balanced (unbalanced) voltage digital interfaces. Multipoint or party-line bus systems require tri-state driver outputs. The more recent RS-485 standard defines this type of electrical connection with up to 32 driver/receiver pairs.

The RS-485 standard specifies that:

- transceivers should be able to withstand a ± 7 V common voltage, regardless of whether power is applied, and have input resistance > 12 kΩ and capacitance < 50 pF
- each driver must be capable of providing a minimum level of 1.5 V when 32 transceivers and two line terminating resistors are connected (for a twisted pair line the terminating resistors will be typically 120 Ω)

2.10 THE EIA RS STANDARDS

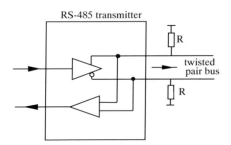

Figure 2.17
Diagrams showing bias resistors (a) used to prevent echo signal spuriously switching the input stage of a transceiver

- receivers must be capable of detecting levels down to 200 mV. (this low level is of great advantage when long line lengths are required)

The combination of long line lengths, high node count (32 maximum), and the use of low cost medium (twisted pairs) has led to the RS-485 specification becoming a widely used data communication standard for industrial applications.

Reflections from electrical discontinuities in a line are a major source of bandwidth limitation in any electrical bus system. Bias resistors (as shown in Figure 2.17) are often used to prevent reflections from causing spurious pulses to be generated by the receiving circuit. The bias resistors effectively prevent the echo signal from entering the switching region of the input stage. Typical bias resistance values are, for 1–10 nodes, 2.7 kΩ; and, for 40 nodes, 27 kΩ. In practice, the amplitude of the reflected wave might be smaller than expected because discontinuities are distributed throughout the line and the resulting echo signals will not be in phase with each other, thus providing a cancelling effect which decreases the magnitude of the reflections. A daisy chain wiring scheme is often recommended to eliminate reflections caused by the connecting stubs used in a multidrop topology.

Knoll and Colo (1992) describe a 2.5 Mb/s Arcnet multidrop network using a 24 AWG solid copper unshielded twisted pair with a characteristic impedance of 120 Ω. They report successful operation with 2.7 kΩ bias resistors at 900 feet with 2 nodes, 800 feet with 6 nodes, 700 feet with 12 nodes, and 600 feet with 16 nodes.

Integrated circuit technology continues to improve the performance of transceiver circuits. For example Texas Instrument claim that their SN75LBC179/180 permits 50 Mb/s operation (with careful system design) with low power consumption (26 mW) and with conformance to the 485 standard.

3
Standards

3.1 INTRODUCTION: Of course we believe in standards, that's why we have so many of them

The majority of the networks discussed in this book are defined by national standards or by well supported (e.g. by trade associations) proprietary standards. The IEC Fieldbus is the only serial network supported by an international activity, but although work on this standard started in 1985, it is still not complete. It is therefore appropriate to consider the place of standards in product development, standards documents and standards organizations before proceeding with a detailed discussion of serial networks and their application. In this chapter a general discussion of the need for standards defining man–machine and machine–machine interfaces is followed by a brief review of the media access methods commonly used in serial networks. The final three sections present a broad introduction to standards, standards documents and standards organizations.

3.2 THE MAN–MACHINE AND MACHINE–MACHINE INTERFACES

Practical application of information technology requires the efficient and economical transmission of data and instructions between a diverse range of spatially separated units. The large number of parallel bus systems that are currently available reflects the importance of this area of system design. Serial data highways are also being increasingly used to interconnect devices and systems. The serial network connection method is used because wiring, installation, extension and maintenance costs are reduced. Applications for serial data highways can be found in process control, factory automation, intelligent buildings and home automation, medical instrumentation, measurement instrumentation, aircraft and military systems. The wide range of applications is reflected by the large scale of the serial network standards activity (see Table 3.1). An internationally accepted serial

Table 3.1
Serial network standards

Serial bus	Application area
Arcnet	Industrial and office automation
ARINC 629	Civil aviation
CAN	Vehicles (also used in industrial automation)
CEbus	Domestic automation
Echelon LON	General automation
FIP	Industrial automation
IEEE 1118 (Bitbus)	Industrial automation
IEEE P1073	Medical instrumentation
IEC Fieldbus	Factory automation, process control and intelligent buildings
MIL-STD-1553B	Defence systems
PROFIBUS	Industrial automation

highway standard is required if the users wish to create multivendor systems is to be satisfied.

It is important to recognize that implementation details are not comprehensively specified by a standard and testing for conformance of a product to a particular standard will be essential. Experience suggests that implementation options are likely to be allowed by any of the serial highway standards and since multivendor systems are a prime objective it is clear that testing for interoperability of equipment (which is not guaranteed by conformance to a standard) will become increasingly important.

The current serial data highway activity is a typical example of a standards-led product development. A widely accepted standard serial highway will dramatically change measurement and control instrumentation, facilitate its inclusion within the framework of information technology, and enable a communication-oriented approach to total automation.

A need for a standardized communications architecture to allow the interconnection and interworking of multivendor information processing systems was recognized internationally when the first attempts were made to produce a general description of the communications process. This work resulted in the open system interconnect (OSI) model (Madron, 1989) which comprehensively defines the task needed to establish an interconnection. This model subdivides the interconnection tasks into a series of layers of specification; each layer is designed to be independent of the layers above and below it, so layers may be changed as technological capabilities change without affecting the operation of the overall system.

Seven layers have been designated for the OSI communications model, and they have been named from layer 7, respectively, as the application, presentation, session, transport, network, datalink and physical layers (Figure 3.1).

3.2 THE MAN-MACHINE AND MACHINE-MACHINE INTERFACES

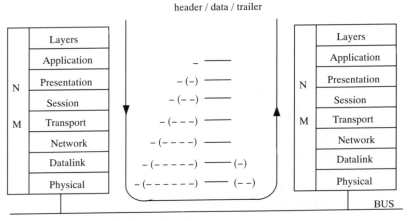

NM = Network management

Figure 3.1
Diagram illustrating the OSI seven-layer communication model

When the data frame is constructed for transmission it is passed successively through layer 7 to layer 1 with each layer adding header and, in the case of the datalink and physical layers, trailer bits. The data frame is reduced to user data by performing the reverse operation of successively stripping header and trailer bits when a received frame progresses from layer 1 to layer 7. A first definition of the model appeared in 1978 (ISO 7498); many additions have been made and these are collectively referred to as the Open System Interconnect (Judge, 1988; Gray, 1991).

In measurement and control the manufacturing automation protocol (MAP) was an early user of the OSI reference model. With MAP each of the layers of the model can use more than one standard, with the result that care is required to ensure that multisupplier systems will operate correctly. The network, datalink and physical layers of MAP are covered by the IEEE 802 family of standards (Madron, 1989). At the field device level implementations of the full seven-layer capability is too complex, too slow and too expensive for many practical instrumentation purposes (Pimentel, 1989). An enhanced performance architecture (EPA) has been devised to allow a seven-layer link to full MAP to be combined with a reduced layer capability that effectively allows the application layer to communicate directly with the datalink and physical layers. Some of the functions of the missing layers must be included in the interface between layers 7 and 2. When used with mini-MAP nodes only having the two bottom layers, the EPA offers a MAP-defined approach to interconnecting field devices. The reduced layer approach (i.e. using layers 7, 2 and 1) has been adopted for the IEC fieldbus standard where the collection of small and time critical data items is important.

The perceived advantages of intelligent instrumentation in new architectural configurations has been a strong motivation for the development of serial data networks. However, the word intelligent must be used with care. Shneiderman (1992, 1993) has noted that the vision of computers as intelligent machines is giving way to one based on the use of predictable and controllable user interfaces. In this case, the computer simply becomes a black box that enables users to accomplish their goals by directly manipulating screen representations of familiar objects and actions. It is interesting to note that natural language interfaces have not progressed beyond the clumsy and slow stage compared with the rapid development of methods for the manipulation of information on high-resolution displays. Users often complain that with intelligent systems the autonomy of the system works against the user, with the result that the user often does not know what the machine is going to do next. Automation systems must be made considerably simpler to operate and configure; they should cater for the needs of the technically oriented user and not just for the information technology specialist. Computer controlled co-operative operation should be the objective of designers. Shneiderman (1993) suggests that labelling machines 'intelligent' limits the imagination of designers, who should have a much greater ambition than to make a computer behave like a super-intelligent human agent.

3.3 MEDIA ACCESS METHODS

A serial network is effectively a highway that is used to transfer information between devices connected to the highway. A device coupling electrical signals to all devices connected to the network should prevent any other devices from transmitting; all of the other devices should be receivers. If more than one device transmits then signals will be corrupted and errors will occur. Several media access methods have been devised, either to ensure that only one device can transmit at any one time, or to provide a mechanism that will allow contention conditions to be detected and corrected.

The master–slave (or command–response) method requires a device connected to the bus to be used as a master device (i.e. a bus controller). Sole control of information transmission on the bus resides with the master device. The master device is responsible for the initiation of all transmissions. Hence no terminal shall transmit unless first invited to by the bus controller and specifically commanded to transmit. This fixed master method is easy to implement. This method is specified by, for example, MIL-STD-1553B. The more complicated token passing method (see below) uses a time-sequenced multiple-master technique.

Unlike the client–server model (where co-operation between pairs of application processes are defined) the producer — distributor — user (PDU) model

3.3 MEDIA ACCESS METHODS

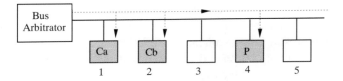

Nodes 1, 2 and 4 recognize variable identifier transmitted by bus arbitrator. They are primed either to transmit (produce) data or accept (receive) data

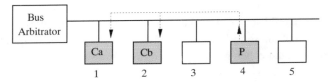

Nodes 1 and 2 receive data produced by node 4

Figure 3.2
Diagram illustrating the producer-distributer-user bus access method

(see Figure 3.2) is based on the broadcasting techniques. In this case the addresses of variable transmitters (and receivers) need not be known by the applications, and most of the data carried by the bus are 'named' variables. The distributor requests producer stations to be broadcast variables but the producer does not know which consumers will use its variables. A consumer uses only the variables it is allowed by the configuration to consume. Clearly the PDU method is similar to the master-slave method in that the distributor acts as a master device. The difference is that the devices require addresses with the master-slave method, whereas the PDU method uses the 'names' of variables to alert producers and users that a bus transaction is required. The factory instrumentation protocol (FIP) use of the PDU method is discussed in Section A.7.

The token used in the token passing method is a controlled frame composed of a unique signalling sequence which is transmitted after each information transfer sequence. The station which holds the token has exclusive right to transmit. This right to transmit may be temporarily donated to the other station for use in acknowledging a transmission by the token holder. To maintain control loop sampled data rates, tokens are timed so that within a specified time interval the token holder relinquishes control of the medium by passing the token to the pre-determined next station. The token is passed from station to station in a circular fashion as if on a logical ring. This token passing sequence is referred to as the token loop. All active stations participate in the token loop. Maintenance of the token loop is implemented by functions within the stations providing for token loop initialization, lost token recovery, additions of new stations to the token loop and reconfiguration when stations are removed from the token loop. Token loop maintenance functions are replicated among all the stations on the network. It should be noted that the station holding the token becomes the bus master, so the token passing method is in effect a master–slave method with each slave taking it in turn to become the master.

The carrier sense multiple access (CSMA) method requires the transmitted signal to be monitored, and if a distorted signal is received this is assumed to be caused by multiple transmissions interacting (i.e. signal collisions). The receiving station issues a retransmission request. Under heavy traffic conditions, requests for retransmission will make the collision situation worse.

The basic CSMA method is clearly not suitable for practical application. A practical system can be designed by arranging for the transmitting station to send a jamming signal when a collision is detected which will be recognized by all the other stations and prevent the operation of their transmitters. The end of the jamming period is monitored by all of the stations and this starts a randomly defined delay period. The station with the shortest delay period will be the first to start transmitting. If a collision occurs again this procedure is repeated and in some cases a longer delay increment is due. After some predefined number of transmissions with collisions occurring, an error condition is reported. Note that this error condition could also be caused by reflections from a break in the cable or by a faulty transmitter being permanently on. This collision detection (CD) extension of the basic CSMA method is called the CSMA/CD method. (A collision avoidance (CA) version of the CSMA method (CSMA/CA) has

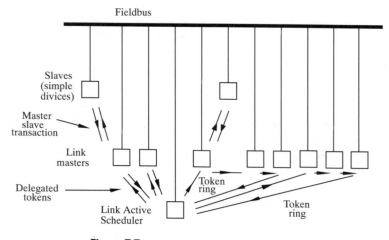

Figure 3.3
Diagram illustrating IEC fieldbus access method

3.3 MEDIA ACCESS METHODS

been developed. This method uses predefined time slots during which each transmitter is permitted to transmit).

One of the proposed IEC fieldbus access methods combines the multimaster token passing method with the PDU access method. A link master (LM) is defined to co-operate with peer entities in connection management. It can also function as a link active scheduler (LAS) which functions as the master station unless a fault condition occurs, in which case another link master could perform the LAS function. The LAS can delegate a token but maintains control of the network through a time-out mechanism. Other LM stations can pass the token (when received from the LAS or from another station) to other peer entities. Two types of token are used; they are called the circulated token (CiT) and the delegated token (DeT). Figure 3.3 shows that the circulated token corresponds

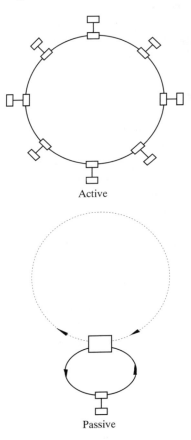

Figure 3.4
Active and passive interfaces to a communicating ring

to the normal token passing function, while the delegated token is sent by the LAS to a master device to transfer the right to perform network transactions for a pre-established period of time. This token does not lead to the formation of a logical ring and can be used for transactions that have to take place at precise instants that cannot be guaranteed by a token passing mechanism. Note that any device holding the token can talk on the network using the PDU access method and that slave devices can be defined as devices that cannot hold the token.

In a ring system, active or passive interface devices can be used to intercept messages. As shown in Figure 3.4 an active interface is part of the ring (so all messages pass through it) while a passive interface is effectively a sub-ring which accepts all messages but they are not returned to the main ring. Access to the ring can be controlled by assigning a time slot to each interface during which it can insert a message, or null bytes of data can be transmitted, and when this is detected by an interface a message can be inserted, or an interface can broadcast messages repetitively. The broadcast method requires a much higher data rate to achieve the same performances as the other two methods.

3.4 STANDARDS

Standards have been said to mirror the progress of industry and the problems brought in its wake. This is certainly true of the measurement and control industry. The scene is one of fundamental change: from standards primarily concerned with the product to those of the process of producing the product; from material aspects to those of interoperability; from pre-occupation with form, fit and function to the embracing of human safety and total quality; from post-product standards to pre-product; from the application of standards as a utilitarian tool to their incorporation as the very essence of a marketing strategy and from national to international standards.

The aims and principles of standards are succinctly stated in British Standard's BS 0, 'A standard for standards'. BS 0 states that standards should be wanted, used, impartial, planned and they should not be duplicated. In addition they should promote: the quality of products, processes and services by defining those features and characteristics that govern their ability to satisfy given needs, i.e. their fitness for purpose; improvement in the quality of life, safety, health and protection of the environment. They should promote the economic use of materials, energy and human resources in the production and exchange of goods. Standards should facilitate clear and unambiguous communication between all interested parties, in a form suitable for reference or quotation in legally binding documents; and they should promote international trade by the removal of barriers caused by differences in national practices.

3.4 STANDARDS

Virtually all measurement and control products involve interconnection with other devices. If they do not now this is likely to change in the future as management insists on the right to access any available information in attempting to drive up gross profit margins. The importance of serial network standards can only increase.

The extent to which a manufacturer can successfully provide products in an early entry position, ahead of standards, can define the market. Initial profitability is high but eventually the finite market size of users prepared to be locked-in is reached. At this point opening of the proprietary standard benefits: the originating manufacturer because the potential market again expands; the other manufacturers because they have access to an installed base; and original users because their investment is protected and they can now choose from a broader range of standards-based suppliers. This is an example of one strategy which demonstrates the power of post-product standards.

As standards emerge, manufacturers must then fight for supremacy through leadership in added value; for example, through rapidly introduced technological innovation and superior quality of their overall service; in other words quality control of the process of transferring added value to the user. It is, perhaps, this dramatic increase in the speed of introduction of new technology achievable through standards that will force the greatest change in business attitudes. A strategic attitude towards standards will be vital. From a relatively dull aside in corporate affairs, standards will become an integral part of marketing strategy since standards will be a key factor in defining the market. Already this maturation of standards from product to process has produced major changes in company attitudes. Corporations are accepting that third party auditing of quality systems (i.e. ISO 9000) is no less logical, no more a loss of 'independence' than financial auditing and demands equal executive attention. This is just as well, since such auditing systems are the catalyst for new levels of overall quality and overall profitability and environmental standards auditing is waiting in the wings.

Post-product standards have the major advantage of being tried and tested, usually through a significant installed base. Inevitably they represent the properties and priorities of their past creators. This may be the recent past but in reality they will not represent the requirements of the new object. They will tend to be supplier-driven as opposed to user-driven, thereby re-enforcing the tendency to represent the past as opposed to the future. In either case, standards based on pre-existing models can result in an excess of variants which spell long-term implementation problems for the user requiring 'open' multivendor systems.

A system is open when it communicates using equipment based on standards that are freely available to everyone. Open systems are most closely identified with the International Standards Organization OSI and with POSIX (the IEEE portable operating system interface definition; Unix is no longer the only POSIX-complaint open system). On a smaller scale several serial network standards can be viewed as open systems despite their proprietary basis; for example, Arcnet and HART are defined by standards supported by trade associations. The IEC Fieldbus will eventually become a major open system for the measurement and control industry.

Proprietary standards can become *de facto* open standards. Microsoft MS DOS and Intels 80386 processor are examples of *de facto* standards. They are owned by the firms that produce them and these firms are in complete control of the standard. The only constraint is market forces. These forces effectively create the *de facto* standards, which then enable rival organizations to produce product clones. However, this presents problems since cloning leads to risks arising from the vagaries of copyright legislation.

A major factor in the creation of pre-product standards is the balance of power between users and providers. The tendency of organizations to increasingly concentrate on their core businesses, together with the technical complexity of the subjects, is already making it difficult to find users who are able to make major contributions to standards work. If future standards are to satisfy the levels of expectation of impartial investors and legislators in terms of availability and applicability, a solution must be found. If this does not emerge from within the existing participants then it may have to be found from outside, because the increasing reliance on standards is unlikely to diminish.

Clearly a business strategy for developing products designed to be used in open system communication environments must involve the establishment of a detailed standards knowledge base. A consequence of products based on open standards is that these products will often benefit from only limited patent protection. Fortunately active participation in the work of relevant standards committees will advance a company's understanding of worldwide market and emerging trends, and provide feedback for product improvement; it is clear that standards development and product innovation are now parallel activities. However it should be noted that the smaller innovative organization will find it difficult to contribute staff effort to standards work since the typical standards committee makes significant demands on the limited expert resources available for this work.

3.5 STANDARDS ORGANIZATIONS

Worldwide standardization is organized by the International Standards Organization (ISO) and the International Electrotechnical Commission (IEC). Figure 3.5 shows the large number of national bodies that supports their work and other relevant standardization organizations. (The abbreviations used in

3.5 STANDARDS ORGANIZATIONS

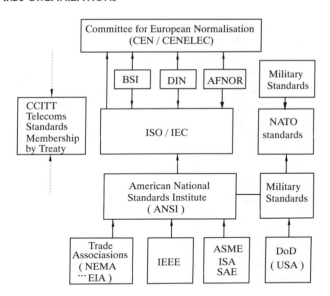

Figure 3.5
Block diagram showing major standards organizations

this figure are as follows: BSI, British Standards Institute; DIN, Deutsche Institut für Normun; AFNOR, Association Française de Normalisation; ISO, International Standards Organization; IEC, International Electrotechnical Commission; NEMA, National Electrical Manufacturers Association; EIA, Electronic Industries Association; IEEE, Institute of Electrical and Electronic Engineers; ASME, American Society of Mechanical Engineers; ISA, Instrument Society of America; SAE, Society of Automotive Engineers; DoD, Department of Defence.)

The ISO is a worldwide federation of national bodies, at present comprising 87 members, one in each country. The scope of the ISO covers standardization in all fields, except electrical and electronic engineering standards, which are the responsibility of the IEC. Together, the ISO and the IEC form the world's largest non-governmental standardization system for voluntary industrial and technical collaboration at the international level. The results of ISO technical work are published in the form of international standards. The ISO published 7438 International Standards in 1990, 8205 in 1991 and 8651 in 1992.

The IEC was founded in 1907 and the ISO was founded 40 years later. In some cases the ISO and IEC have overlapping areas of interest. A Joint Technical Committee (the first one to be formed was JTC1) has been formed to address all aspects of the information technology field. It is intended that this committee will reduce duplication of effort and contention between IEC and ISO members in the areas of information systems and microprocessor systems.

In Europe the European Committee for Standardization (CEN) deals with general standardization work whereas the European Committee for Electrotechnical Standardization (CENELEC) deals with standards broadly relating to electrical engineering. They are non-profit technical organizations composed of the National Electrotechnical Committees of 18 countries in Western Europe. Together they form the Institute for Standardization in Europe (Camp, 1992).

CENELEC is chartered to develop a coherent set of voluntary electrotechnical standards as a basis for creating a single European market without internal frontiers for goods and services inside Western Europe. CENELEC work will override the work of any European national group. This work is based primarily on the international work of the IEC. (In 1992 87% of CENELEC standards were adopted from IEC standards). CENELEC is working to harmonize the European standards activity by establishing European standards (ENs), Harmonization Documents (HDs) and European Pre-Standards (ENVs). National organizations such as the British Standards Institute (BSI), the Association Française de Normalisation (AFNOR) and Deutsches Institut für Normung (DIN) support the work of CENELEC and the IEC.

An American national standard is a document which has been adopted by the American National Standards Institute (ANSI) after verification that a national consensus exists to favour the national endorsement of the document as an American National Standard. There are now more than 11 000 American National Standards and more than 1000 new standards are added each year. National organizations such as trade associations e.g. the Electronic Industries Association, (EIA), the Institute of Electrical and Electronic Engineers (IEEE), the Instrument Society of America (ISA) and the American Society of Mechanical Engineers (ASME)) support the work of the American National Standards Institute.

Since standards are based on the work of a committee of experts it would not be surprising if some of these experts hold intellectual property rights (IPR) that could affect the success of the standard. The IPR would normally be protected by a patent. Patents held by inventors not directly involved with the standardization process could also affect the success of the standard. Clearly standards organizations must develop administrative procedures to deal with conflicts of interest. For example, if the ANSI receives notice that a proposed standard may require the use of a patented invention then a published procedure is followed. Basically ANSI does not object in principle to drafting a standard that includes the use of a patented item, provided that sound technical reasons justify this approach.

A large number of certification bodies have been established to check that a product is in conformance with a particular standard. If conformance is established then products may be appropriately marked (In the UK the British

Standards Institute uses the Kite mark and the European mark is the CE mark). The European Organization for Testing and Certification (EOTC) was established in 1992 to co-ordinate the work of the certification bodies. Organizations have been established to validate the work of the certification bodies (for example, in the UK this is the National Accreditation Council for Certification Bodies).

3.6 THE STANDARD DOCUMENT

A good serial network standard should facilitate the interoperability of implementations of the standard. A good standard should satisfy the following requirements:

- the desired functions must be defined clearly
- the standard should contain no superfluous specifications
- options should not affect interworking
- it must be possible to make implementations conforming to the standard
- implementations must be unambiguously testable.

Using natural language, it is difficult to make clear specifications which cannot be misinterpreted. Tuinenburg (1990) proposes that a standard should be defined in three parts:

- the definitive standard, defined in a formal language that can be used to prove that no inconsistencies are present
- a natural language version which includes implementation guidance
- specifications of tests that can be used to assess the conformance of an implementation to the standard. These tests will be designed using the formal language definition of the standard.

The standards described in Appendix A.1 are primarily natural language documents. In some cases state machine chart notation is used to define the detailed operation of a data link layer. The IEC Fieldbus standard is a notable example where the need to use formal techniques (and at least state machine charts) to define the function appears to have been completely ignored. One of the driving forces for the development of the IEC standard was the need to offer serial network products that would interoperate. It is unfortunate that the IEC standard has a large number of options that will lead to sub-set conformance and mitigate against interoperation.

It is important to recognize that a test for conformance cannot be guaranteed. If all tests indicate conformance, this is in fact an inconclusive result. A test for non-conformance can, however, be used with confidence; a failure of just one test indicates non-conformance with absolute certainty. A bad standard will make unrealistic (and ambiguous) demands of implementations and introduce non-testable specifications. Irrelevant and superfluous specifications will waste testing time and add nothing to the functionality of implementations.

Products manufactured to be in conformance to a particular technical standard will also be required to satisfy other standards, for example, safety and EMC (electromagnetic compatibility) standards. In addition the introduction of the EC Product Liability Directive has emphasized the importance of manufacturing products to conform to national and/or international standards (Sumner, 1991). The EC directives are driven by a concern for the rights of the consumer and with a concern for the quality of the environment. They are having a considerable impact on product design and manufacture. For example, the EC directive on EMC requires manufacturers to fulfil two basic requirements, namely; apparatus must be constructed so that any disturbance that it generates allows radio and telecommunications equipment and other apparatus to operate as intended; and apparatus must be constructed to provide an adequate level of protection against external disturbances. To satisfy these requirements manufacturers must either manufacture in conformity with specified European standards, or draw-up a technical construction file which describes the apparatus, sets out the conformity procedures to be adopted, and includes a technical report or certificate from a competent body. The manufacturer must certify that the apparatus complies with the directive by making a declaration of conformity. This must be kept available for inspection by enforcement authorities for ten years following the placing of the apparatus on the market.

Farrell (1990) notes the many reports of allegations that vested interests often slow down or even prevent the completion of a standard. This is especially so when there is integrated circuits or complete products already available in the market. Most industry committees operate under (and are delayed by) the so-called consensus principle which requires that the final form of the standard document is supported by unanimity (near unanimity); or they at least require that all views will be respected (this leads to multiple options) and are not over-ridden by the majority. Changes to these rules would help but are slow to be accepted. (The European union recently changed the rules under which EC standards are set: unanimity is no longer required). Considering the diversity of the bodies contributing to the IEC Fieldbus document, it is not surprising that progress has been slow and many options have been included in this standard. Any attempt to accelerate the standardization process will lead

3.6 THE STANDARD DOCUMENT

to equally serious concerns about collusion among any vendors who may be actively working to complete the standard. A widely accepted international serial network standard will not be easy to achieve.

Standards can clearly be of considerable relevance to the success of a manufacturing organization. It is important that designers of serial network products should be aware of the standards implications of their work.

4
Factory Automation and Process Control

4.1 INTRODUCTION

Automation is necessary to satisfy modern quality specifications and it is required to satisfy, for example, the requirements of flexible manufacturing, just-in-time production and environmental legislation. It provides the means for the economical use of raw materials and energy. Automation involves more than just the application of negative feedback principles to achieve closed-loop autonomous operation of a process. Substituting closed-loop control for manual operation of a plant has a long history but it is only recently that semiconductor technology has provided the components that enable programmability and communication capabilities to be integrated and applied at relatively low cost to the automation of industrial processes. The trend towards larger and more complex plants, necessary to take advantage of economies of scale, have been largely facilitated by advances in control, integrated circuit, software and communication technology.

The distinction between process automation and manufacturing automation is diminishing. Process automation is concerned with the automatic operation of continuous processes such as, power plant, rolling mills, cement kilns and petrochemical plants. These processes have very large material flows and often have stringent safety requirements. Manufacturing (factory) automation is concerned with the production of discrete objects. Initially this involved isolated production islands for maximum quantity throughput, and involved numerically controlled machine tools with punched paper tape programs. It was not until much later that these islands were networked to achieve flexible manufacturing, with short re-tooling times and low inventories. Process and manufacturing automation activities have traditionally developed as separate activities. However they now use similar electronic systems for closed-loop control, for operator–machine interfaces (and visualization software) and for networking. This convergence of the two activities is emphasized by the IEC Fieldbus standard which has been designed for process and factory automation applications.

4 FACTORY AUTOMATION AND PROCESS CONTROL

4.2 PROCESS CONTROL

Signal communication standards have been at the very core of measurement and control's major evolutionary steps. Modern process measurement and control instrumentation date from the early 1950s. At this time instrumentation was predominantly mechanical with pneumatic signalling using 3–15 lb/in^2 as the preferred signal range. Standards at this time were post-product, and the first reference to 3–15 lb/in^2 in British Standards appeared in 1962. From the mid 1950s, 4–20 mA analogue signalling over twisted pairs of copper wire, connected in a star or point-to-point configuration had spread to become an industry norm. The fact that BS3586 (the 4–20 mA standard) was published in 1970 illustrates the long delay that is characteristic of post-product standardization.

The first international standard in the process field was published by the IEC Technical Committee (TC-65) in 1971, namely IEC 381-1 which established the 4–20 mA signalling range as an international standard. This was the year following the appearance of BS 3586, but from 1973 the BSI began to give preference to international work. This was also the year that CENELEC was formed and from then on BSI implemented many IEC standards without change as dual numbered British Standards.

Standardization began to be attempted in advance of the physical realization of the product. Work on PROWAY, a process system serial communication standard, began within IEC in 1975, and a sub-committee of TC-65 was set-up in 1981 to provide a focus for this work. Also in 1981, international standards work began on programmable controllers. By 1983 work was underway on safe systems, by 1985 on the fieldbus, and by 1987 on the functional safety of programmable electronic systems. Nevertheless, despite the scale and importance of this work, proprietary standards continued (and still continue) to appear and satisfy user needs.

Hammond and King (1980) discussed the advantages of a hierarchical structure to control large complex processes. Figure 4.1 gives a block diagram representation of this type of system. The evolution of the hierarchical system structure is discussed by Papovic and Bhatkar (1990). Depending on the speed of the process under control, d.d.c (direct digital control) software could be multiplexed around several control loops. As microelectronic technology developed relay-based sequential controllers were replaced by programmable logic controllers. Note also that connections to the plant were point-to-point and consequently the cost of installing and configuring cables was significant.

Cabling costs can be significantly reduced by eliminating the need for long lines between the controller and its associated sensor and actuator. A block diagram representing this form of distributed system is shown in Figure 4.2. This architecture was pioneered by Honeywell in 1975; their TDC 2000 system

4.2 PROCESS CONTROL

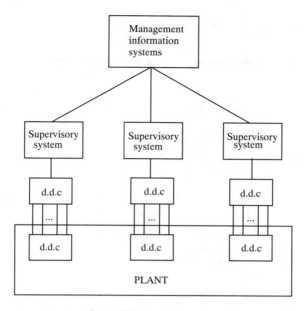

Figure 4.1
Hierarchical process control

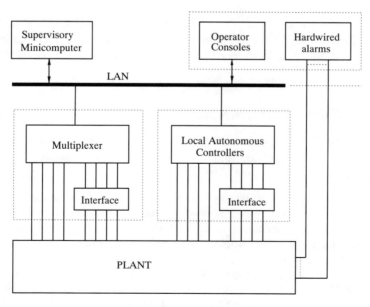

Figure 4.2
Distributed process control

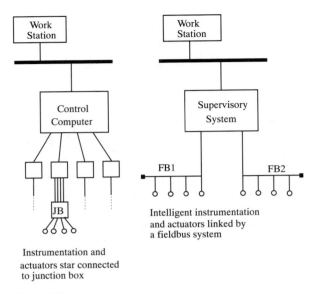

Figure 4.3
Comparison of star-connected instrumentation with the fieldbus connection method

was the first distributed system for continuous plant control. Other manufacturers followed this lead and produced their own systems (Papovic and Bhatkar, 1990; Kalani, 1988). A further reduction in cabling costs is achieved by interconnecting field-located devices with a serial bus (the fieldbus as shown in Figure 4.3).

4.3 FACTORY AUTOMATION

Hammond and King (1980) noted that as long ago as 1965, Coales in his inaugural address to the IEE Control and Automation Division made the following observation.

> 10 years ago, we thought that by now we should have many automatic factories in which the plant was controlled by digital computers to produce goods to customers orders according to the most efficient programme possible. It was expected that by now there would be a superfluity of consumer goods and excessive leisure for all, even if many would be out of work. It was this picture that was given by the popular Press and made many think in terms of Frankenstein. They could have been spared their anxiety, because, although in the past 10 years there have been enormous advances in the application of automation, and greatly increased productivity, the pushbutton factory is still almost as far away as it was in 1955, and we still have an acute shortage of labour.

4.3 FACTORY AUTOMATION

Hammond and King (1980) continued as follows.

> There are still no truly automatic factories, at least in the UK; but all the indications are that the combination of our current theoretical and practical knowledge of how to design such factories coupled with the availability of new computational tools based on microelectronics could combine to ensure that industrial operations are very different 10 years hence from their heavily manned state today.

A decade later it is possible to confirm the validity of these predictions. The next significant development in automation equipment, namely field-located intelligent instrumentation and actuators linked by a fieldbus defined by an international standard, is well underway. It can be confidently predicted that this will have a large impact on the nature and scale of employment in the manufacturing industries.

The distinguishing feature of factory automation is that it involves discrete parts manufacturing rather than continuous flow processes. Manufacturing operations are grouped into cells. Figure 4.4 shows the block diagram of a typical machine tool cell using three axis control systems. If necessary work may be transferred between cells by automatic guided vehicles or a robot system and in this case local microcomputer-based controllers store and execute

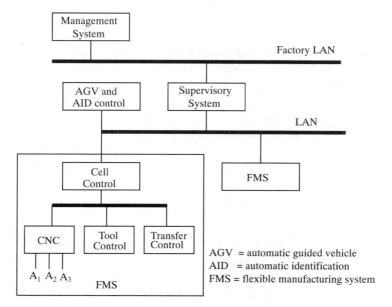

Figure 4.4
A typical machine tool cell

measurements as required. Work in progress can be identified by the use of product tags which are interrogated by a cell system when a new work item arrives for processing. In general, mechanical systems used in discrete parts manufacturing are much faster than process control operations and it is often essential to establish autonomous local control loops to satisfy speed requirements. An objective of modern automation systems is to provide flexible solutions to manufacturing problems. Flexibility is obtained by a combination of software design and the use of fieldbus techniques to simplify installation and configuration, and to facilitate the addition and removal of field devices.

In 1980 the American company General Motors established a group to develop an independent computer network capable of sustaining a multivendor equipment environment on the factory shop floor. This work has resulted in the creation of the Manufacturing Automation Protocol (MAP) and by 1984 the first specification (version 1.0) was published. An international MAP activity has been established and by 1987 version 3.0 of the MAP specification had been published (Dwyer and Ioannou, 1989). A Technical and Office Protocol (TOP) is being devised for office and design applications and it has the same methodological base as MAP. MAP and TOP are both based on the OSI seven-layer reference model (Morgan, 1987). At each of the layers of the model more than one standard can be used, with the result that care is still required to ensure that multisupplier systems will operate correctly. The network, data link and physical layers are covered by the IEEE 802 family of standards (Madron, 1989).

The full MAP communication spine is a broadband 10 Mb/s system. It is multichannel, and access to the network is by the token-passing method. Carrier band MAP is single-channel, slower (5 Mb/s) and has a lower cost. The carrier band is suitable for linking together clusters (cells) of instrumentation. The typical topology of a MAP system is shown in Figure 4.5. At the device level implementation of the full seven-layer capability is too complex, too slow and too expensive, and a method is required to send short messages (with a modest data overhead) at frequent intervals. Several devices have been specified to enable communication networks to be interconnected. The most complex requirements arise when an OSI network is to be connected to a non-OSI network. The gateway architecture uses the full seven-layer stack to achieve this connection.

Communication using a gateway is slow because every transaction is translated via its full seven layers. The bridge unit links two compatible networks through a common data-link layer, hence networks linked must use the same addressing scheme and frame size. Since the bridge uses only layers 1 and 2 and they are transparent to the transmitting node, the communication rate is reduced by its use. The bridge would be used, for example, to link a broadband MAP network with MAP carrier band subnetworks. A router links two

4.3 FACTORY AUTOMATION

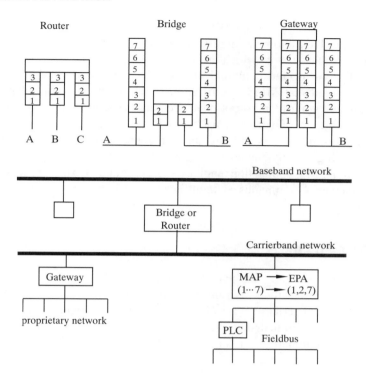

Figure 4.5
Topology of the Manufacturing Automation Protocol

or more networks with a common network layer protocol and enables different networks with a common network protocol to be linked. The router needs to be addressed; it is not transparent to transmitting nodes. The existence of these inter-networking devices underlines the problems of establishing a universal standard approach to communication networks. Since user requirements are so varied it is likely that a need for these devices will always exist. The use of proprietary networks exacerbates this problem. It is clearly desirable that designers of future equipment should avoid specialized non-standard networks.

The enhanced performance architecture (EPA) and the mini-MAP nodes provide a solution to the problem of providing MAP-related communication for instrumentation within a cluster of equipment. An EPA node implements a seven-layer link to full MAP and in addition allows an application to communicate directly with the data link layer. The mini-MAP node implements only the bottom two layers and needs to be linked to an EPA node to communicate with the full MAP system. Equipment manufacturers are now providing proprietary units to connect field equipment to MAP systems. For example, the MAP Equaliser (Reflex Manufacturing Systems Ltd, Crawley, West Sussex, UK) allows a large range of shop floor devices, such as PLCs, machine tools and robots which do not have MAP interfaces, to be connected to a MAP network and communicated with via the MAP MMS standard. It should be noted that MMS specifications cover numerical control machines, programmable controllers, robot controllers and process control systems.

The emergence of software packages based on the MAP Manufacturing Message Specification is making an important contribution to its success. MMS is a high-level language to establish communications between MAP devices. It allows application engineers to program a MAP network using easily understood commands, without needing to know the details of network operation.

4.4 THE FIELDBUS

It is important to recognize that the International Standard fieldbus will not instantly replace installed field instrumentation systems. In addition, although a large number of serial networks have been defined for field applications, they have not, as yet, eliminated star networks using 4–20 mA analogue signalling. Hence an evolutionary path for the adoption of the fieldbus is likely to be followed, as shown here.

- Replacement of analogue transmission, over existing wires, by a superimposed digital signal. The use of low-quality cable and long home runs will lead to the use of low bit rates, e.g. 31.25 kbit/s.

- In practice star networks are implemented as a trunk and spur topology, with a cable tray carrying the bundle of home run connections (the trunk) to a central control room. The home run trunk connectors can be replaced by a fieldbus, and relatively short spur connections, using existing wiring, can be connected to the bus in a junction box.

- New plant will use mixed media (including electrical twisted pair, fibre optic and radio links) for the fieldbus to make best use of available technology. The lower 4 mA of the 4–20 mA signalling method provides power to field devices. Hence a fieldbus replacement should be designed to allow the supply of power with the communication signals. For operation in hazardous areas the field device and its power supply must be designed to satisfy intrinsic safety requirements. In addition the fieldbus must be designed to comply with EMI and EMC regulations (IEC 801 (1 to 5)).

The topology finally adopted will be constrained by the need to isolate potential fault conditions and the need to keep all devices of each control loop on the

4.4 THE FIELDBUS

same fieldbus. If too many separate fieldbus systems are used the wiring and installation benefits of the fieldbus will be reduced. Hence it will be necessary to ensure that each fieldbus segment is as large as possible, consistent with the above fault limitation constraint. It is important that the fieldbus protocol supports reliable and timely transfers of messages among different fieldbus segments, and should allow a connected device to be addressed regardless of its location (segment) and physical connection method.

It is clear that many functions traditionally performed by higher-level control systems will eventually be distributed down to field devices. The impact this will have on control room design has received little attention. It could result in a simplification of control room operations (which will be beneficial) but the safety implications of installing complex programmable electronic systems throughout a plant will need careful consideration. Although the fieldbus appears to be at the lower level of the industrial automation, control and communications, hierarchy, it is capable of absorbing the higher levels of intelligence and operating in a self-contained manner which will leave the higher level systems to perform a supervisory role.

The concept of time criticality is the feature of the fieldbus (or any other field serial network) which distinguishes it from the local area networks (LANs) used at higher levels in the automation hierarchy. A fieldbus system should guarantee information flow between field devices and between these devices and higher level systems, subject to time constraints imposed by the dynamics of the process to be controlled. The time criticality constraints arise from monitoring requirements and, particularly, the need to support real-time control. The timing performance of a serial network can be described by the delays introduced by the message passing through the levels of the protocol and by propagation delays introduced by the physical media.

The fieldbus communication protocol must be designed to support cyclic (periodic) transactions and event-driven aperiodic transactions. The fieldbus will also be required to support non-critical-time event-driven transactions. Clearly regular cyclic data transactions and unrestricted event-driven transactions can conflict, especially when the bandwidth of the bus is heavily used. To overcome this problem the fieldbus should be designed to allow the user to establish a privileged position for some transactions and assign levels of priority to each event-driven transaction. The addition (or removal) of a field device should not impair the existing time critical behaviour.

Fieldbus communication services which support periodic update or on-demand update can be used to convey synchronization commands. These commands enable events to be accurately time-stamped and clock synchronization to be achieved across the system.

The fieldbus should be designed to guarantee a given quality of the transferred information. This is usually referred to as Quality of Service (QoS). In the case of time-non-critical transactions QoS is expressed in terms of correctness and completeness of the transferred information without any time constraint. For time-critical transactions QoS is expressed in terms of correct, complete and timely reception (i.e. available in a defined time window) of information by all relevant application processes. Application processes must be informed of QoS status for each transaction. The services and quality of service offered by a fieldbus system should be user-selectable to satisfy application requirements.

Typical quality of service features are listed below:

- unsatisfactory transaction detected and reported
- unsatisfactory transaction detected, and application processes exchange common data to check for consistency (timeliness, value and, in the case of data lists, order)
- overload of time slots allocated to specific transactions

These quality of service features are related to the level of system integrity available for the fieldbus user to specify. Typical features of this type include: system operation with no undetected errors; any faulty device should not stop fieldbus communication between other devices; redundancy of key operational functions; and bus duplication for full network redundancy.

The fieldbus should be designed to enable the possibility of changing, adding and removing devices without upsetting transactions between other devices. Furthermore it is important that failure of any device should not propagate to upset the overall fieldbus operation. Data type and length transported by the bus should be user-defined and not predefined by a bus standard. They will be set-up by the network management functions at the configuration stage before the bus is used in its operational mode. Service functions (configuring and monitoring) should be possible through a non-vendor-specific tool which may or may not be permanently attached to the fieldbus. Users of fieldbus systems expect that the cost of attaching field devices will be low. Users also expect the electrical drawings, design and maintenance activity associated with field installations to be considerably reduced when the fieldbus approach is adopted.

Network management is clearly an important aspect of any fieldbus system. Many of the features discussed above will be provided via the network management function. Services provide by a fieldbus system can usefully be partitioned into two groups. One being concerned with the messages and data conveyed between field devices, and the other being concerned with the overall behaviour of the communication network. In practice the communications behaviour of a network will be tested and verified before it is put into use as a fieldbus. During normal operation users will not have access to the service setting network management functions.

4.4 THE FIELDBUS

A fieldbus designed for both process control and factory automation applications will be expected to satisfy a wide range of operating requirements. The length of the bus, number of devices connected to a bus and the update rate will depend on the application, which can range from process control in the chemical industry to discrete parts manufacturing in the electronics industry.

For chemical industry applications a bus length in the range of 1000–2000 m will be required with branches in the range 10–20 m. Approximately 30 devices may be connected to a single bus and a complete plant could use several (10 to 100) fieldbus systems. In less safety-conscious process applications approximately 200 devices could be connected to a single bus. Power over the bus (with intrinsic safety) will be required as an option.

Relatively large plants are used by the chemical industry and very large time constants can be expected. Hence it is expected that typical bus transactions involving five data bytes per device will use an update rate of four times per second.

In some cases it will be important to protect the integrity of particular control loops by dedicating a fieldbus to the sensor, actuator and indicator of each loop. Provision for interfacing a hand-held terminal to each bus should be provided to enable technical support staff to check bus and device operation. Hence it is possible that only four devices will be connected to the fieldbus to maintain single loop integrity.

The hand-held field interface with the bus will become increasingly important as the level of intelligence in the field sensors and actuators increases. More information will be transferred to and from these intelligent devices and therefore a higher data rate will be required. At this stage it is difficult to assess the data rate implications of intelligent devices but it is likely that a rate higher than 250 kbit/s (probably 1 Mb/s) will be appropriate.

Factory automation using faster mechanical operations will require higher update rates. The automotive industry uses cell-manufacturing techniques. In this case a bus will typically serve PLCs with a fast bus with 32 devices connected with five data bytes transferred at an update period of 10 ms. The discrete parts manufacturing techniques used by the electronics industry involves small lightweight mechanisms operating at relatively high speeds. In this case a typical scan period could be 8 ms for 32 connected devices with one to eight data bytes per message.

International fieldbus activities started in the mid-1980s with the formation of a fieldbus working group by the International Electrotechnical Commission (IEC) (Wood, 1990). The IEC and national groups (e.g. the SP50 committee of the Instrument Society of America (ISA) and the AMT7 committee of the British Standards Institute) have brought together a complete spectrum of industrial experience to bear on the problem of specifying a serial data highway standard for measurement and control. Unfortunately a consensus has been found to be difficult to obtain, with the different philosophical bases of central media access control and token passing control being particularly difficult areas for agreement. It is not surprising that the currently published draft standard contains a significant proportion of options for a product conforming to the standard. This will make the construction of multivendor systems difficult despite the existence of an international standard.

National groups in France (FIP, the factory instrumentation protocol project) and in Germany (PROFIBUS) have developed national fieldbus standards. Both of these projects have progressed to a stage well in advance of the IEC standard with protocol circuits and fieldbus configuration software available for product investigations. Fortunately FIP and PROFIBUS support the work of the IEC, so the long-term prospects for an international fieldbus standard are still encouraging.

The FIP fieldbus uses the broadcasting operating principle. A bus arbitrator, which may be implemented in different stations to provide redundancy, organizes the broadcasting function. The broadcasting method offers the configuration advantage that the address of transmitters and receivers need not be known by the applications. A message producer does now know which are its consumers, and the consumer inputs only the variables allowed by configuration to consume. Cyclic traffic where the arbitrator sends names (i.e. from a table) periodically, and aperiodic traffic where the free time slots in the cyclic traffic can be used, are possible operating modes. Aperiodic traffic can result in the arbitrator calling for variables or by the arbitrator allowing the station to transmit a message to another station or to all stations (broadcast). The FIP standard is defined by layers 1, 2 and 7 of the OSI communication model.

PROFIBUS uses the master–slave (command/response) access method. It also uses a multimaster concept that allows a logical token ring to be established to link the masters. A master holding the token is permitted to communicate with the passive slave devices that are also connected to the bus. The bus access method, as well as the data transmission and management services, are based on that defined by ISO 8802.4 and the provisions of the IEC 955 standard Process Data Highway, Type C (PROWAY C). The PROFIBUS standard is defined by layers 1, 2 and 7 of the OSI communication model. In addition the specification of layer 7 has been designed to offer automation systems facilities for communication with field devices. This has been achieved by creating a subset of the ISO layer 7 MMS protocol specified by MAP.

A number of other, standard defined, serial networks, for example Arcnet, Bitbus, Echelon LonWorks and the controller area network (CAN), are in use in industrial control applications. Arcnet was developed by the Datapoint Corporation in the late 1970s. It was one of the first networks used to interconnect personal computers. It has now been used in a wide range of industrial and office applications. However it was not until 1992 that an

4.4 THE FIELDBUS

ANSI specification committee wrote an Arcnet protocol standard. In 1984 Intel responded to the need for a low-cost method to interconnect microcontrollers by developing Bitbus. It has been used in a wide range of applications and was soon incorporated into an IEEE standard. IEEE Std 1118 (1990) expands upon Bitbus without obsoleting existing devices. The Echelon local operating network (LON) is the most recent addition to the fieldbus scene. The Echelon Corporation was founded in 1986 to offer a complete off-the-shelf solution for designing and implementing measurement and control networks. Echelon claim that, whereas the IEC fieldbus is merely a communications standard, they are providing users with a complete system that incorporates a communications standard. This is a proprietary standard and it is an open architecture in the sense that low-cost protocol circuits (the Neuron chip) are available and any devices using these circuits will be able to communicate with each other. The controller area network (CAN) was developed by Bosch for use as a vehicle serial bus and is therefore especially useful for fast response communication over short distances. Wide acceptance of the CAN protocol by the automotive industry and the availability of integrated circuit implementations is leading to its use in the complete range of measurement and control applications. All of these networks are supported by integrated circuit supplied by the major semiconductor manufacturers. However only the Echelon network offers the high-level software support necessary to construct a fully operational system.

The Arcnet network (Conner, 1988) uses a token passing protocol at a 2.5 Mbps data rate and supports up to 255 network nodes with data packet sizes up to 512 bytes. Low-cost coaxial cable, twisted pairs and fibre optic links can be used to form the network. The high cost of implementing the original system has been considerably reduced by the introduction of the Arcnet protocol integrated circuit, manufactured by the Standard Microsystems Corporation. It can now be used in embedded control applications typically found in factory automation and process control. An Arcnet LAN standard is published by the Arcnet Trade Association. In a token passing communication network each node is assigned a unique node identification number (1 to 255 in the case of the Arcnet protocol). Each node passes the token to the node with the next highest node identification number but may transmit only when it has possession of the token. When a node receives a token it transmits data onto the network before passing on the token, and if a node has no message to transmit it simply passes on the token. To illustrate the timing capability of Arcnet, consider a particular node on a 10 node network; it receives the token every 23.51 ms, assuming an 1 μs propagation delay between nodes and assuming that each node transmits 508 bytes of data.

The Bitbus communication protocol, a central media access technique, is a modified version of the IBM synchronous data link control (SDLC). The IEEE 1118 standard is based on HDLC (high-level data link control), an ISO version

(ISO 3309) of the bit-oriented synchronous protocol. This allows the use of one master node and up to 250 slave nodes. Each node has a microcontroller which arbitrates control and interface function. Slave nodes cannot indicate transfers on the bus: they only respond to instructions from the master device. The physical connection of devices is defined by RS 485 and can be used in a synchronous or self-clocked connection mode. The synchronous mode uses two twisted pairs, one to carry the clock and one to carry the data (using this method it is possible to interconnect up to 28 nodes over 30 m at data rates between 500 kbps and 2.4 Mbps). If repeaters are not used, the asynchronous self-clocked connection mode requires only the data pair allowing communications over longer distances. Bus systems based on 1118 can be configured as a single-level hierarchical network with a master controlling several slave units. A multiple level hierarchy is constructed by arranging for the slaves to have an associated submaster which is used to control another multidrop bus.

The basic philosophy of the Echelon network is that data should be broadcast only when they change with the result that distributed systems will be data-driven rather than command driven (Rabbie, 1992). For systems with critical timing requirements it is possible to arrange for nodes to be polled. It has been designed for the full range of media: twisted pair, power line, radio, infrared, co-axial cable and fibre optics. An RS 485 twisted pair operates up to 625 kb/s and a transformer coupled twisted pair operates up to 1.2 Mb/s. The protocol used is based on the OSI communications model and it is implemented by a special-purpose integrated circuit called the Neuron chip. The Neuron chip offers a full seven-layer version of the communications model. Echelon's protocol (LonTalk) is based on a modified version of the carrier sense multiple access (CSMA), media access control (MAC) algorithm. This effectively enhances the MAC functions of the data link layer improving performance for multiple-media communication, sustaining performance during heavy loads and allowing large networks to be managed. A collision detection mechanism controls the use of the bus.

The controller area network (CAN) operates with bit rates up to 1 Mb/s with a maximum latency of 150 μs for highest priority messages at the maximum bit rate. It is a connectionless network and its multimaster mode of operation will allow any node to send data when the bus is free. Several error control mechanisms are offered, including cyclic redundancy check (CRC). Repeat transmission is initiated when errors are detected and defective nodes are automatically located and disconnected. Bus contention is avoided by the use of dominant bit patterns which are related to the priority status of the message. This bus has been designed for short message (up to eight bytes per frame) transmission, over short distances, used in real-time control application.

Each of the serial networks discussed here has been adopted for use in a variety of commercially available measurement and control systems. For

4.4 THE FIELDBUS

example the Broderson System 2000 is based on the Bitbus communication protocol and offers 375 kb/s (with a 300 m limit on the cable and no repeater) and 62.5 kb/s (with a 1200 m limit on the cable and no repeater. It uses RS 485 signalling over twisted pairs to link a maximum of 27 slaves, without repeater, and 250 slaves with repeaters. Overall system control is achieved with two data base levels; a local data base at each slave and a common data base in the master containing an exact copy of all of the slave data bases. Information transfer is achieved by copying information from one data base to the other. Slaves are polled cyclically by the master to update its data base. Slave data bases are updated by a local scan operation. At power-up the Bitbus master searches the system by polling all possible slave addresses and uses the configuration information, stored in each slave at power-up, to build a copy of the slave data base in the master. When power is first supplied to a slave, it automatically checks the number and type of modules communicating with the slave and builds its data base accordingly. If slaves are changed the master must be reinitialized. Error control is based on the use of 16 bit CRC and timer-controlled polling retry.

In an attempt to satisfy market requirements some manufacturers are offering more than one bus system in their product range. One of the National Standards with one or more from Arcnet, Bitbus and CAN is a common arrangement. In addition an interface to the VME parallel bus system is typically provided and this provides an interface to a LAN at a higher level in the automation hierarchy.

The main features of the IEC Fieldbus electrical physical layer specification are the use of the multidrop bus connection for up to 32 devices with twisted pair cables carrying asynchronous data transmission using Manchester coding and half-duplex communication. A voltage mode (with high and low speed options: 2.5 Mb/s, 1 Mb/s and 31.25 kb/s) and a current mode (1 Mb/s) is specified. A clip-on transformer connection is specified for the current mode bus with a maximum cable length of 750 m. The current mode bus uses a 20 kHz current-fed sine wave to supply power to remote devices and the 1 MHz digital data signal is added to the sine waveform. With the eventual inclusion of fibre optic and radio links, it is clear that a large number of options will be a feature of this standard. The IEC/ISA fieldbus standard committees working on the data link layer have reconciled the two fundamentally different control methods of token passing and bus arbitration. A compromise approach has been adopted which includes bus arbitration functionality in a token passing bus. The bus arbitrator approach was championed by the French national group but its popularity was established originally by MIL-STD-1553B. The problem of options was, and remains a major concern with the fieldbus data link protocol definition. This will result in a minimally conforming implementation not being suitable for use in all possible circumstances. The IEC data link standard has progressed to the point where silicon implementations of the standard can be manufactured.

Kirk and Wood (1994) present a broad overview of the IEC Fieldbus standard and list the eight parts of the standard; namely

- IEC DIS 1158(pt. 1), Fieldbus — Introductory Guide
- IEC DIS 1158(pt. 2), Fieldbus — Physical Layer Specification
- IEC DIS 1158(pt. 3), Fieldbus — Data Link Service Definition
- IEC DIS 1158(pt. 4), Fieldbus — Data Link Protocol Specification
- IEC DIS 1158(pt 5), Fieldbus — Application Service Definition
- IEC DIS 1158(pt. 6), Fieldbus — Application Protocol Definition
- IEC DIS 1158(pt. 7), Fieldbus — Fieldbus Management
- IEC DIS 1158(pt. 8), Fieldbus — Conformance Testing.

Part 2, The Physical Layer Specification, is approved and Parts 4 and 5 are in the voting stage. The remainder are under development.

A function block or device description language is required to describe common application tasks with standard parameters and compatible interfaces. The Instrument Society of America SP50 working group has completed a considerable body of work in this area. This work is currently not within the field of interest of the IEC Fieldbus Committee.

4.5 INTELLIGENT TRANSDUCERS

A sensor (S) is an input device that responds to some physical property (involving radiant, mechanical, thermal, electrical, magnetic and chemical processes) and generates an electrical signal which can be measured. When combined with interface electronic circuits (I) the sensor becomes an input transducer (T_{in}) as shown in the block diagram in Figure 4.6. A transducer accepts energy of one form and converts it to energy of another form. A transducer is constructed to be directly available for application by an end-user, hence it is packaged appropriately with compensation, calibration and signal conditioning circuits.

An output device accepts a signal (usually electrical) and produces output energy of another form. Typical output devices are; LEDs, loudspeakers and hydraulic rams. Actuators are a special class of output device that produce a higher energy output involving a mechanical movement (force), e.g. DC/AC motors, stopper motors, electromagnets, pneumatic and hydraulic rams, and piezo-electric stacks. Combination of output devices with interface electronic

4.5 INTELLIGENT TRANSDUCERS

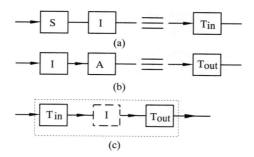

Figure 4.6
Block diagram showing the formation of transducers from the basic sensor and actuator

circuits creates another form of transducer (an output transducer, T_{out}) as shown in Figure 4.6.

The simplest general situation is for an input transducer to drive directly an output transducer. Often signal (and resistance) levels will be incompatible and additional interface components will be required, as shown in Figure 4.6. Note that the block diagram representation usually does not show the excitation input required by a particular device. This is often a convenient omission that simplifies diagrams, but it hides the fact that an output is effectively obtained by the modulation of an external power supply. Further important omissions from these diagrams are the spurious noise generators that all devices effectively contain and the additional input that temperature of the environment (or internal electrical components) will provide.

The use of the word intelligent does not imply the intelligence level of the human operator. A very low level of intelligence will qualify a device to be called an intelligent device. A device commonly becomes an intelligent device when it automatically performs functions that were previously implemented by a human operator, e.g. ranging and calibration, and decision-taking associated with the communication and utilization of data. Smart or autonomous devices are alternative names for the intelligent device.

An intelligent field device will have one or more of the following features.

- Automatic calibration and ranging under the control of a built-in digital system. Calibration constants may be automatically acquired and stored in the local memory of the field device. Autoconfiguration and verification that the hardware is operating correctly should follow internal self-checks.

- Automatic correction of offsets, time and temperature drifts, and non-linear transfer characteristics.

4 FACTORY AUTOMATION AND PROCESS CONTROL

- Self-tuning (or adaptive) control algorithms (inclusion of fuzzy logic control may obviate the need for this feature).
- Different control programmes may be down-loaded from a host system (or from the local store in to field device) to enable dynamic reconfiguration of control functions.
- Control may be implemented via a serial bus and a host system, directly via the bus or directly by special connections to the field devices (see Figure 4.7).
- Condition monitoring for fault diagnosis. This may require additional sensors in a field device and may involve digital signal processing, and data analysis software.
- Communication circuits for remote interfacing via a serial bus. (A hand-held communication unit could be interfaced via the bus or by means of a special

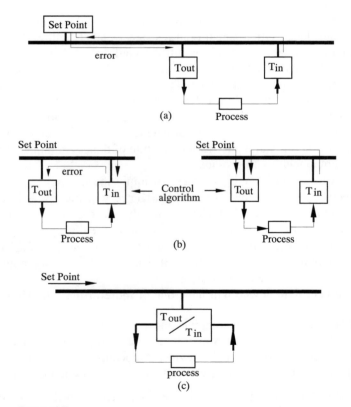

Figure 4.7
Diagram showing three methods for forming a closed loop using a serial bus to connect field devices

4.5 INTELLIGENT TRANSDUCERS

interface to the field device using optical, radio or inductive techniques. They will obviate the need for external operator controls on the field device).
- Depending on the processing and memory capability of the field device summaries of acquired data can be sent over the bus in engineering or scientific units.
- Testable at manufacture or in use via the bus.

The main features of the intelligent transducer (incorporating sensors or actuators) are shown in Figure 4.8, and Figure 4.9 shows in greater detail the essential units required to implement an intelligent sensor and an intelligent actuator. Devices that do not contain a microprocessor but can interface with one are no longer considered to be intelligent devices. The ability to communicate over a serial bus is the most recent distinguishing feature of an intelligent device.

Several implementation possibilities can be identified.

- A separate sensor with several integrated circuits mounted on a PCB (or thin/thick film substrate). An integrated circuit interconnect bus, e.g. the I^2C bus, will often be used. The total chip count will depend on, for example, the way the microcomputer is realized (e.g. the number of memory chips used) and on the quality (and complexity) of the analogue interface components.
- If the market is sufficiently large, ASIC techniques could be used to reduce the chip count. For example a separate sensor could be supported by an

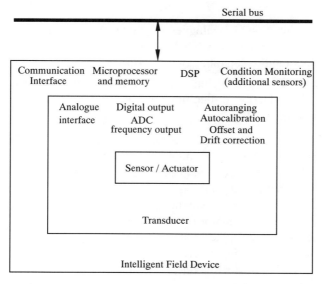

Figure 4.8
Diagram showing the main features of an intelligent field device

4 FACTORY AUTOMATION AND PROCESS CONTROL

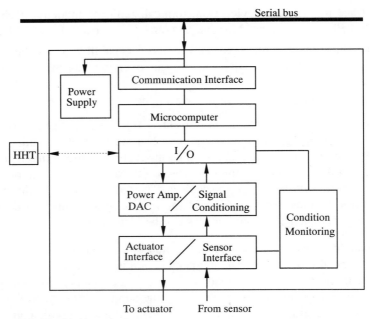

HHT = hand-held terminal

Figure 4.9
Block diagram showing the major components of an intelligent sensor and an intelligent actuator

ASIC implementing all analogue and digital functions required to interface the sensor to a communication integrated circuit. Note that power supply components require separate consideration.

- For very large markets a single ASIC solution may be appropriate. Exceptionally, if production difficulties are not too severe, the sensor could be included on the ASIC.

Clearly a detailed knowledge of market conditions, manufacturing techniques and the circuit capability (and availability) of ASIC technology is required to make the best implementation choices for a particular product.

The major activity in the intelligent devices field is in sensors to measure pressure, acceleration, displacement (proximity) and temperature. Pressure sensors account for the majority of the activity with approximately 75% of the world market. Displacement accounts for approximately 50% of the remaining market, and temperature and acceleration each account for 50% of the remaining 12.5% of the market.

World revenue forecasts for the intelligent sensor market area indicate that at the beginning of this decade it was about $400 million dollars and at the end of this decade it is expected to be well in excess of $1000 M dollars. It is interesting to note that the regional share (USA, Europe, Japan) of the market is about 60:20:20. Harmer (1991) has presented a review of European research in advanced sensors.

A consequence of including high-level system functions in the field device is that bus traffic could be increased by, for example, condition monitoring (fault detection) and configuration data configuring new devices to a distributed system). Although the typical small message format of fieldbus systems leads to low bus utilization it is important to design an intelligent field device so that it does not unnecessarily increase bus traffic.

4.6 POWER OVER THE BUS

Current loops have been widely used for analogue signalling. The live zero 4–20 mA signalling range allows the lower 4 mA to be used as a power supply for remote devices. In addition, voltage levels can be restricted and protective components added to allow remote devices to be operated in hazardous environments when intrinsic safety requirements must be satisfied.

The Highway Addressable Remote Transducer (HART) protocol was designed to make direct use of 4–20 mA equipment designed to operate with a superimposed modulation enabling digital communication to be established between a remote device and a host system. It provides a method to use existing point-to-point wiring (from a junction box) and it allows hybrid use of equipment where analogue signalling is mixed with digital signalling or the link can be all digital with the analogue capability used to supply only the power to remote devices. Devices designed to use the HART protocol can be connected in parallel, and an addressing procedure allows each unit to set its analogue output to 4 mA (which acts as the power supply) and forces the device to communicate only digitally. This parallel connection effectively converts the twisted pair into a multidrop bus. The HART specification limits the maximum number of multidropped devices to 15, and therefore the maximum current supplied by the power source will be 60 mA. A useful feature of this approach is that it allows industry-standard power supply methods to be used. The HART protocol (which is master–slave with a data rate of 1.2 kb/s) is discussed in more detail in Section A.1.8.

A similar current-fed approach to supplying power to remote devices has been developed by Solartron (Schlumberger Instrument Division). In this case the current levels are much higher since the remote devices are more-complicated, field-located, multichannel, data collection and processing devices, which act as connection points for local sensors and actuators. The serial transmission network (S-Net) supplies power and digital communications to isolated measurement pods (IMPs) with a power requirement less than 1 W per Imp. Up to 50 multichannel stations (Imps) can be interconnected and linked to a host computer. A greater number can be connected by using additional addressing information. S-Net uses a polling protocol at a data rate of 163 kb/s. It is claimed that the network protocol ensures that up to 1000 data channels can be triggered and the results returned to the host computer in less than 1 s.

Burton, Stone and Kirk (1988) have described options for power transmission in their 1553 bus 1988 proposal to the ISA SP-50 committee. Low-frequency AC and DC voltage supplies were considered. Since the 1553 bus uses transformers to couple the 1 Mb/s data rate digital signal, it is necessary to provide a parallel path around each transformer for low-frequency power supply signals. The feasibility study undertaken as part of the project associated with the ERA proposal concluded that MIL-STD-1553B could form the basis of an intrinsically safe, 2 km bus, with power supplied to a limited number (less than 5) terminals via the bus cable. Successful field trials have demonstrated the practical validity of this work.

The Centre for Industrial Research, Oslo, Norway (SI: Senter for Industriforskning) has developed a radically different approach to supplying power to remote devices (ISIbus, 1990). They use a current-sourced bus with a data rate of 1 Mb/s. The digital communication signal is superimposed on a current-fed 14 MHz sine wave which is used to supply power to remote devices via a transformer coupling technique. By using ferrite E cores it is possible to use the bus cable itself as a single turn primary which couples with a multiturn secondary (with of the order of 20 turns). The secondary voltage is rectified and smoothed to form the power supply for a remote device and the much higher frequency communication signal is obtained by high-pass filtering the transformer-coupled signal. An attractive feature of this approach is that it is possible to design the transformer coupler so that it can be easily clipped on to the twisted pair serial bus. When the 14 kHz power supply current is set to 100 mA (r.m.s) the voltage drop across each coupled device is typically 1 V (r.m.s) when 100 mw is delivered to each device. The maximum number of nodes is 32 for both a powered and a non-powered bus. An important feature of this power supply method is that a faulty field device (i.e. with a short or open circuit connection) will have no effect on the operation of the rest of the network.

For operation in hazardous areas, barriers located in the fieldbus power supply limit the available power to 1.5 W. The number of devices that this power level will drive depends on the power taken by each device and the cable specification. An alternative approach would be to locate the barriers with the couplers, and in this case the cable must be fully protected. It should be noted that current industrial practice is to locate the barrier outside the hazardous area. Attempts to distribute barriers in the hazardous area will require specification consideration by safety certification organizations.

4.7 SAFE SYSTEMS

Serial networks enable the construction of distributed systems using field-located programmable (intelligent) electronic systems. The field locations often need intrinsic safety design requirements to be satisfied to prevent energy in the field device exceeding a threshold that will cause an explosion to occur. Intrinsic safety is discussed in Chapter 2. This section introduces the design implications of a system entering an unsafe operating mode due to failures of sub-units in the system.

Tolerable risk level is a key factor in the safety integrity requirement specification of a safe system. Risk is defined as the combination of the frequency, or probability of a hazardous event, with the consequences of that event. Any risk must be reduced as far as reasonably possible to a level which is as low as reasonably practicable. This is referred to as the ALARP principle.

The performance of the safety system in carrying out its safety functions must be characterized. In practice a safety integrity parameter is defined and called the system integrity level. Safety integrity applies solely to safety-related systems and it gives the likelihood of a safety-related system satisfactorily performing the required safety functions. International standards are under development that will state tolerable risk levels for specific applications and state the required system integrity levels for safety-related systems designed for specific applications (IEC, 1992, and other application-specific standards).

Safety protection systems are usually designed to be separate from the basic process control system (BPCS). This requirement for safe systems was noted by the UK Health and Safety Executive in 1987 (UK HSE, 1987). More recently the health and safety requirement of the harmonized European standard EN 292 (CEN, 1991) invalidates the use of a so-called simple system, irrespective of the technology, to perform both control and safety functions. In practice a separate emergency shutdown (ESD) system is used which typically monitors parameters such as water levels, pressures of liquids and gases, temperatures, and the position of valves and fire check doors. Fire and gas (F & G) systems are often designed as separate systems: they use dedicated field devices to continuously monitor smoke, temperatures and gas (toxic and combustible) levels. When a threshold condition is exceeded these systems will generate an alarm indication and automatically initiate action to return the plant to a safe state. Often the speed of response will be faster than could be consistently expected of human operators; for example a small ESD system (e.g. with less than 1000 field devices) will typically complete its shutdown actions in less than 50 ms. Shutdown systems must be reliable, both to protect plant and personnel, and to minimize the time (downtime) for which a process is not operating.

A safe system will be characterized by the following features:

- separate control and protection systems (at low safety integrity levels some safety functions can be included with the BPCS but common mode effects soon become a major problem)
- design to eliminate common mode events
- design to use components of known reliability which have operated in similar applications
- high-quality power supplies
- designed to minimize the effect of electrical interference
- man–machine interface designed to respect the data-handling capabilities of the human operator
- conservatively designed serial networks that will not overload or unacceptably delay the transfer of alarm signals
- software designed with good documentation and quality assurance procedures.

In contrast to the BPCS (which is continuously operating) the protection system is inactive until an alarm condition occurs. Consequently, without a system testing methodology, failures may occur which will not be detected by the protection system. Practical experience has shown that by applying fault-tolerant design techniques (e.g. as used by the aerospace industry) protection systems with duplicated or triplicated units can tolerate faults without loss of service. Modern protection systems use microcomputers to continuously perform self-diagnostics. A large microcomputer-based protection system (MBPS) will clearly be a complex system, but nevertheless, for example, an MBPS involving 5000 field devices can respond automatically to a hazard condition in less than one second.

Redundancy has been extensively investigated by the space, aerospace and process control industries. Many papers have been published. Kirrman (1987), for example, has reviewed the use of fault tolerance techniques in process control. He notes that there is no general-purpose fault-tolerant architecture, and that only a thorough knowledge of the plant controlled by the fault-tolerant system can lead to a cost-effective architecture. Several levels of redundancy have been investigated. For example, triple modular redundancy, the 2-out-of-3 (2oo3) structure, uses three operating units and detects failure from the results of a pair-wise comparison. When a failure is detected, operation can revert to a 1oo2 operating structure and the failed channel is repaired. With 1oo2 structures the comparison operation detects a failure but an internal self-test arrangement is required in each channel to detect the failed unit. However this type of redundancy cannot be easily applied to the typically quiescent ESD or

4.7 SAFE SYSTEMS

F & G system. Recent work in this area indicates that the most economical and effective solution uses two dual systems. In this case one system is configured as a control system using 1oo2 voting with the objective of maintaining the plant on-line and within specified operating parameters. The other system uses a 1oo2 voting scheme designed to provide personnel and environmental safety. This approach attempts to satisfy the overall system objectives of maintaining availability and ensuring safety.

Software for large process control and factory automation systems can have extremely complex source code. When serial networked intelligent instrumentation is fully established a significant proportion of this software will no doubt be distributed to the field devices, but nevertheless the overall supervising system will still require complex source code. In addition the interaction of the distributed software systems will introduce further problems that may lead to unacceptable safety performance.

Often the space of possible operating modes for the software is too large for it to be tested in a reasonable time scale. To confirm that a failure rate of 10^{-9} failures per hour has been achieved it is necessary to test a programme for many multiples of 10^9 hours. In addition programmers are unable to visualize all possible operating conditions, so the software is incomplete and failures caused by this type of error are difficult to identify and eradicate.

Littlewood and Strigini (1992) discuss the results of an investigation of software reliability, which showed that about a third of all error conditions found were such that they produced a failure about once in 5000 years of execution. After correction of the more significant errors only the 5000 year errors will remain to make the system unreliable, and elimination of one of these will bring negligible improvement in reliability.

Testing for correct operation is difficult and any results purporting to establish the reliability of a system should be rejected if the test methods used are not defined in considerable detail (Parnas, Schouwen and Kwan, 1990). It is necessary to ensure that test inputs causing failures are encountered with the same frequency with which they would in fact arise in practice. In addition, deciding whether an input is correct is itself a difficult problem. Clearly a simple observation of a programmes behaviour over a short time scale provides little information. The software testing problem is closely related to the software design problem.

In recent years a considerable research activity has been directed towards using mathematical proofs to guarantee that a program will function according to a specification. These formal techniques (Potter, Sinclair and Tell, 1991) have achieved some success but the problem remains of constructing the specification in terms of formal statements that can be unambiguously processed by mathematical operations. The design of the specification and its conversion to the formal statements representation introduces an error-prone link into the design chain. The completeness and accuracy of the original specification will always be a cause for concern.

Many practical systems have shown the value of hardware redundancy. Software designers can use redundancy concepts, and this typically involves arranging for different software design teams to develop substantially different versions of the software from the original specification. These different versions are run in parallel and a comparison and voting method is used to establish if an error condition has occurred. Ensuring that the design teams adopt completely different design methods is, of course, a major problem.

A system design approach is required that does not make too heavy demands on system software reliability (Leveson, 1991). In this case a sufficiently modest requirement would be one that specifies a reliability that can be demonstrated in a reasonably short time scale. Littlewood and Strigini (1992) close their paper by saying '... we should remain wary of any dramatic claims of reliability. Considering the levels of complexity that software has made possible, we believe being sceptical is the safest course of action'. Clearly, when safety is a requirement, a conservative approach should be adopted for the design of complex measurement and control systems involving serial networked programmable electronic circuits.

5
Laboratory and Medical Automation

5.1 INTRODUCTION

The laboratory and medical automation application areas have been combined in this chapter, but this should not be taken to imply that technology from other application areas can be easily transferred for use in medical applications. The medical user group has definite requirements which must be satisfied by the design of the user interface, and patient safety and tolerance of technology are factors requiring special consideration.

Laboratory instrumentation (which will generally make extensive use of parallel bus techniques) could be used as a controller for a fieldbus system. Also, the use of special bus systems to interconnect laboratory equipment is increasing, e.g. in the case of field-located analytic instruments used in the process control industry.

Several of the serial bus systems discussed in this book could be used in medical applications. However only the IEEE medical information bus (MIB) has been designed to include an application layer specifically designed for medical applications. This standard has been in development for almost a decade but it has still not progressed beyond the status of a provisional standard.

This chapter presents a brief review of the interconnection requirements of laboratory instrumentation. A discussion of the medical information bus follows and the broader systems implications of this bus are established by including a general review of communications networks in health service applications. Considerable advances can be expected in both the laboratory and medical automation application areas.

5.2 LABORATORY INSTRUMENTATION

Laboratory instrumentation is often expensive and typically involves complex electronic systems. VLSI circuit technology is reducing the cost of this instrumentation. It also enables complexity (and optional features) to increase and this is leading to an increasing requirement for computer control. Apart from the increased usability that results from computer control of these instruments, the automation of experiments is greatly facilitated by a bus interconnect method. Large blocks of data are often collected and significant signal (data) processing operations are commonly performed within each instrument. Parallel digital bus connections have been the usual way to interconnect these instruments. Parallel to serial (and vice versa) converters enable serial techniques to be used for long links. High data rate fibre optic serial links for instrumentation applications are currently receiving considerable attention, and it can be expected that they will compete strongly with the parallel bus.

High-energy physics experiments provided an early motivation for the development of bus connected instrumentation. Data read-out and data acquisition systems in particle physics experiments now always use some kind of digital data highway. Internationally recognized standards, such as CAMAC and FASTBUS, developed from the needs of researchers working in the physics area. Also, industrially developed bus systems, such as VME, have become widely used in physics instrumentation. These standards specify a parallel bus with attendant connectors (mounted on a backplane) and contact assignments, signal and timing specification, data transfer protocols, in some cases software routines, and in addition the complete range of mechanical features.

The Computer Automated Measurement and Control (CAMAC) standard was developed in the mid-1960s for applications in high-energy nuclear physics and nuclear energy establishments. A 24 bit parallel bus (with a total of 86 connection wires) carries data and control signals around a unit (called a crate) into which 24 modules and a crate controller may be plugged. Up to seven crates may be connected in parallel via a 66-way bus extension. The CAMAC serial highway system uses a twisted pair connection to link up to 62 crates over longer distances up to 15 m. Much longer distance transmissions are achieved with fibre optic links. CAMAC is specified by IEEE, IEC and BSI standards (BS 5554 (1978), BS 5836 (1980), IEEE 583, 595, 596, 675, 683, 726 and 758). CAMAC has found wide acceptance in nuclear research and other fields, in particular in industrial control. A wide range of modules complying with the standard are still commercially available.

At the time when CAMAC was specified there were no microprocessors and no other bus standards. When the microprocessor appeared the parallel bus standards started to be developed. As the microprocessor capability developed through 8 bit systems to 16 bit systems, FASTBUS appeared and this has now evolved into a definitive ANSI/IEEE specification (IEEE 960). It became an IEC standard in 1990 (IEC 935). In addition the Motorola 68000 microprocessor family became widely used and the suitability of this processor family for use in bus-oriented systems was recognized by the publication and commercialization

5.2 LABORATORY INSTRUMENTATION

of the VME ANSI/IEEE standard (IEEE 1014 and IEC 821). Di Giacano (1990) presents a full discussion of the wide range of parallel bus systems.

Bus data rates continue to increase. The relatively recent scalable coherent interface (SCI) provides bus services by transmitting packets on a collection of point-to-point multidirectional links. Its protocols support cache coherence in a distributed shared memory multiprocessor system, with message passing, I/O and LAN communication taking place over fibre optic or wire links (Gustavson, 1992). SCI uses point-to-point links to achieve very high speed communication. For the highest performance over short distances (e.g. within a cabinet), 16 bit wide links operate at 1000 Mbytes/s. For I/O applications within a room, serial co-axial cable links run at 1000 Mbps, while for I/Os over distances of a few kilometres an optical fibre can carry the same serial bit streams. The scalable coherent interface is defined by an IEEE standard (IEEE1596).

The general-purpose interface bus (GPIB) was introduced by Hewlett-Packard in the mid-1970s for use as a parallel interface for programmable electronic instruments. This parallel bus has 16 wires, of which 8 are for data, 3 for hand-shaking and 5 for control purposes. Transmission at 1 MHz is possible over 15 m. Units to serially extend (with fibre optic and with coaxial cable) are available for communication over distances up to 1000 m with data rates of 50 kb/s. This bus was originally specified by IEEE Std 488 (1975, revised 1978) (Loughry, 1978). It is now also specified by the following IEC standards: IEC Std 625 pt 1 (1979): electrical, mechanical and functional details; and IEC Std 625 pt 2 (1980): coding and formats. Careful initial design has led to the widespread successful international use of the GPIB.

Devices connected to the GPIB must be able to perform one of three functions:

- listener: a device capable of receiving data over the interface when addressed
- talker: a device capable of transmitting data over the interface when addressed
- controller: a device to manage communication by sending addresses and commands.

Devices able to talk, and listen, and devices able to talk, listen and control are allowed, but there can be only one active controller. A maximum of 15 devices may be connected in one contiguous bus. This bus has been widely adopted by manufacturers of electronic test equipment. A typical configuration would include a digital voltmeter, programmable power supplies, a digital oscilloscope and programmable signal generators all coupled to a PC via the GPIB.

The VME standard defines a backplane interface that simplifies integration of data processing and control of peripheral devices (Freer, 1987). It uses the

IEC Eurocard specification to define the boards plugged into the backplane. The VME backplane makes provision for four buses, namely, the data transfer bus, the priority interrupt bus, the arbitration bus and the utility bus. The data transfer bus allows bus masters to direct the transfer of data between themselves and slave units; it consists of 32 data lines, 32 address lines, 6 address modifier lines and 5 control lines. The priority interrupt bus consists of seven interrupt request lines, one interrupt acknowledge line and an interrupt acknowledge daisy chain. Signals that enable periodic timing and coordinate the power-up and power-down sequences of the VME bus system are provided by the six-line utility bus. The arbitration bus allows an arbiter module and several requester modules to co-ordinate use of the data transfer bus; it consists of four bus request lines, four bus grant lines, a bus clear line and a bus busy line. Two signalling protocols are used on the VME bus; one is an open loop protocol which uses broadcast bus signals and the other is a closed loop protocol that uses interlocked bus signals (these co-ordinate internal functions of the VME bus system, as opposed to interacting with external stimuli). The VME bus standard clearly defines a complex parallel bus system. It is widely used and, as shown in Figure 5.1, it provides an important link between serial field bus instrumentation and higher-level system functions.

The VME extension for instrumentation bus (VXIbus) is a relatively new instrumentation standard that takes advantage of the best technologies from GPIB instruments, modular plug-in instrumentation boards and modern computers (ERA, 1990). The VXI bus is an instrument-on-a-card standard that evolved to satisfy the need for sophisticated instruments mounted on

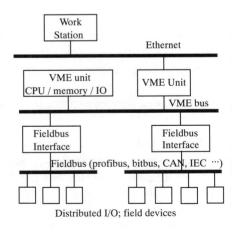

Figure 5.1
VME system linking fieldbus instrumentation to a higher-level system LAN

plug-in cards. A consortium of instrument suppliers introduced the VXIbus specification as an industry standard in 1987. VXI, like GPIB, is backed by a formal specification that ensures an open multivendor platform. Users can programme VXI instruments using high-level message strings, familiar to GPIB users; VXI fully uses IEEE 488.2 and the Standard Committee for Programmable Instruments (SCPI) to ensure ease of programming for complex instruments and a high degree of conformity in the command sets of different instruments. VXI provides a flexible, software-controlled platform for users to acquire data from fieldbus devices and perform data analysis for immediate presentation or for passing on to a higher-level supervisory system.

5.3 MEDICAL INSTRUMENTATION

The use of telecommunication and information technology is central to the efficient provision of a health service. It enables, for example, investigation, monitoring and management of patients to be carried out regardless of location. This involves a combination of topics from the fields of medicine, telecommunications, informatics, and measurement and control; it is referred to as telemedicine. A variety of networks are used, ranging from the ordinary telephone network to special-purpose networks. The Integrated Services Digital Network (Griffiths, 1990) provides a single serial network that is able to satisfy the requirements of a wide range of telemedicine applications.

The smaller-scale communication problems experienced within a health facility are not so well developed. This is despite the technology push of new, and more reliable, sensors and data processing equipment, and the continuing need to make more efficient use of medical personnel. Figure 5.2 shows the major features of a health care system. Telemedicine developments will no doubt improve the link between GPs and patients and the centralized hospital facilities but little has been done, as yet, to take advantage of serial networked medical instrumentation within these facilities.

The UK NHS has stipulated that medical systems must comply with the Open Standards Interconnect. This requires the many different suppliers of medical instrumentation to agree on a standard medical information transaction format and message protocol. The IEEE Medical Data Interchange Committee is working on this problem. This committee aims to specify and establish a robust and flexible communication standard for the exchange of data between heterogeneous healthcare information systems. Their standard, IEEE P1157, is expected to cover all of healthcare communications including that within a healthcare centre, between centres and between individual medical personnel and their central facility.

5 LABORATORY AND MEDICAL AUTOMATION

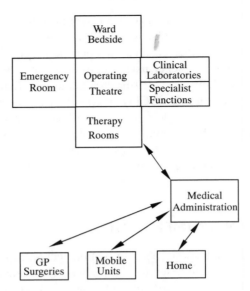

Figure 5.2
Sites where serial networked measurement and control instrumentation could be used in healthcare applications (note that there are approximately 2.5×10^6 hospital beds in Europe (0.4×10^6 in the UK) and 1.3×10^6 in the USA)

It is clear that a wide range of sensors (and actuators) will be required, with an equally wide range of performance features. Consider first the centralized hospital-based facilities. Bedside monitoring involves a range of sensors, for example, for temperature, blood pressure, heart rate and blood gases. The trend here is towards comprehensive monitoring using minimally invasive sensors. Current concern with contact with blood is leading to the development of multi-parameter sensors (e.g. blood gas and pH) in a single catheter. In emergency rooms the particular requirement is for high speed of response and disposable sensors. Operating room surgical procedures require sensors to monitor vital functions such as heart rate, blood pressure, respiration and blood oxygen. Clinical laboratories use a wide range of measuring instruments, many of which involve sophisticated medical automation to load samples. The 'wired' hospital presents an opportunity for a new approach to healthcare management by putting resource utilization on a more informed and factual basis (Wilkins, 1990).

Decentralized facility locations include GP surgeries, ambulance units and patients in the home environment. The telephone network is likely to be the common medium that allows the GP to interact with hospital facilities and

5.3 MEDICAL INSTRUMENTATION

enables data from patients in the home or in transit to be transmitted to a GP or a hospital. Interaction with, for example, closed-loop controlled drug infusion mechanisms would also be possible. The emphasis here will be on portable low-cost instruments. The same standardized network and communication protocols should be used for both centralized and decentralized measurement systems.

There is great potential for the application of closed-loop computer control technology in patient care (Packer, 1990). However it should be recognized that a major problem here is the possibility (however small) of patient injury attributable to the automation system. The fear of the expense of the consequent litigation is leading medical practitioners to ignore its established advantages. These advantages include the increased frequency of correction of drug infusion rates (which can reduce drug usage) and the reduction of routine repetitive tasks that will free nursing staff to devote more time to patient care. A widely accepted standard serial network linking these closed-loop systems to a supervisory system will considerably improve the reliability and acceptability of these systems, particularly if the safe system methodology, discussed in Section 4.7, is used to design the software and hardware.

A local network for bedside-related measurements and procedures has been under development since 1982 (Franklin and Ostler, 1989; and Figler and Stead, 1990). In 1984 an official committee of the IEEE was formed to develop a standard defining a medical information bus (MIB). This has become IEEE P1073 (it is in fact a set of three standards). It is making slow progress to develop beyond the draft standard stage (Gardner *et al.*, 1992).

The MIB committee addressed the problem that many instruments depend on the use of parameters derived from another device and it was therefore necessary to manually input data from one device to another. This was obviously too complex, time-consuming and error-prone. To solve this problem, manufacturers could integrate multiple functions within a single monitor that is associated with a particular patient or users could attempt to integrate monitors from different vendors. The first method restricts the users choice of monitoring equipment. The second method is time-consuming and very costly as there is no economy of scale (each interface must be individually designed). Clearly a MIB could provide a cost-effective, reliable and easy-to-use interconnection method. However it must be defined by a standard, and users must require instrumentation vendors to implement the standard interface in their equipment.

A study of user needs led to the following basic MIB requirements.

- For safety reasons it is essential that devices automatically report their location at a specific bedside as soon as they are attached to the bus.
- For safety reasons each device should continuously display a positive visual indication of communication with a host computer.

- Devices automatically identify themselves as to type and capability when attached to the bus.
- Bedside monitoring devices can produce large amounts of data. A method of filtering this data is required to prevent an overload of the storage capability of the host system.

Using these requirements as guidelines the following MIB objectives were produced.

- Medical devices should be interfaced with a host computer in a compatible vendor-independent fashion.
- High reliability in terms of both transmission accuracy and data delivery as well as network availability, serviceability and survivability.
- Provision of a network that is appropriate for the acute patient care setting, e.g. frequent network configuration and change in equipment location.
- Provision of a simple non-technical user interface.
- Ability to support a wide range of network topologies.
- Cost-effectiveness.

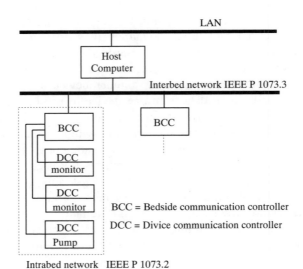

Figure 5.3
The MIB interbed and intrabed networks

5.3 MEDICAL INSTRUMENTATION

The MIB committee made a commitment to use existing standards whenever possible in order to speed-up and simplify the development of the MIB. In particular, the International Standards Organization (OSI) communication model was incorporated into the MIB standard. To support and maintain flexibility the MIB standard is presented as a family of three standards. IEEE 1073.1 deals with the general architecture and components that must exist throughout an MIB implementation, and IEEE 1073.2 specifies the bedside communication subnet. The third part IEEE 1073.3 describes the access function between a bedside communication controller and a host computer located on a conventional LAN. The MIB is effectively an information pipeline that connects a medical device to a host computer. It is composed of two networks, as shown in Figure 5.3, one of which is an interbed multidrop network and the other an intrabed star topology network. The interface between the medical device and the MIB is called a device communications controller (DCC) which connects to a bedside communication controller (BCC) forming a subnetwork. The BCC functions as a convertor between the bedside medical device and the host interface node commonly called the MIB communications controller (MCC). The MIB standard is discussed in more detail in Section A.11.

6
Intelligent Buildings

6.1 INTRODUCTION

Instrumentation and appliances used in domestic applications are supplied by a very large number of distinct manufacturers. Before serial bus control can be established for home automation purposes it is clear that manufacturers must agree to supply devices that will interface with the bus and operate in conformance with a serial bus standard. In addition software must be supplied by a system integration organization that will provide a user-friendly interface for the home owner. As discussed in Section 6.2, it is clear that the take up of serial bus controlled home automation will be a slow evolutionary process. Section 6.2 introduces the factors controlling the use of a standard serial bus for home systems and then discusses the standards that are currently under development.

In commercial buildings the wiring problem associated with the installation of office equipment strongly motivates the development and acceptance of the use of serial bus techniques. The intelligent building has progressed beyond the concept stage to the practical realization of a large number of highly automated buildings around the world. Section 6.3 emphasizes the value of a systems engineering approach to the intelligent building and discusses the major sub-systems. In addition the international standards activity in this area is discussed and the wireless approach to system interconnect using UHF techniques is introduced. A review of serial networks standards used in intelligent buildings is presented in Section 6.4.

6.2 DOMESTIC BUILDINGS

The 'technology push' to develop electronic consumer products for home automation applications is very strong. However when the location of these products is considered (namely the housing stock) then it is clear that the large-scale success of these products will depend on the user environment as well as their technical quality. It is interesting to observe that, even after 100 years of continuous industrial development, approximately 5% of the housing stock in the UK is in poor condition (0.75×10^6 council and 2.75×10^6 private houses) with nearly a million houses being unsuitable for human habitation. Clearly a considerable level of financial support will be required to create a significant number of houses with the construction features that will enable home automation systems to be established. Northcott (1991) notes that although increasingly many microelectronically controlled products are being used, by 2010 it will still probably be only a minority of affluent enthusiasts who go to the trouble and expense of installing fully-integrated house systems for entertainment, security and control of appliances.

Although millions of houses are apparently targets for home automation it is the new house that will be of particular interest. Unfortunately the potential market size is then considerably reduced. For example in the UK private sector house-building has been around 175 000 completions a year since 1965, while public sector housing has fallen from about 175 000 in 1965 to only 25 000 in 1990. Of course not every household will be interested in high levels of home automation and this will further reduce the size of the potential market. (in the UK the significant household types are one person (25%), and married couples (25%) and married couples with children (30%)). It is then clear that the path to home automation will be evolutionary rather than revolutionary. A sub-system will be bought for security, entertainment, lighting and appliance control, and a specialized system will be developed for those who have difficulty using current products. A full agreement on communication standards will be required to provide users with the opportunity of creating multivendor systems.

The need to reduce electricity consumption is a driving force for the adoption of home automation techniques. Users save money if less electricity is used but also CO_2 emissions associated with energy production will be reduced. Since the cost of generating electricity can vary by an order of magnitude over a 24 hour period, a method for limiting demand when it is expensive would be advantageous. Table 6.1 shows how electricity is used in the UK home.

From this table it will be seen that space heating accounts for a large percentage of the total energy consumption. Additional insulation and draught-proofing will reduce this consumption, but designing the heating system to be responsive to environmental changes will also reduce the large energy penalty that will follow from allowing overheating to occur; increasing the temperature of a house by 1°C increases energy consumption by about 8%. A control system that responds to sensors monitoring external environment conditions and sensors that monitor internal conditions will be required to achieve an energy sensitive performance acceptable to the house occupier. A serial bus system will be the most effective way to interconnect the components of this system. In addition the bus will allow finer control of the operation of energy-consuming appliances. For example consider the domestic lighting system: of

6.2 DOMESTIC BUILDINGS

Table 6.1
UK annual use of electricity (10^9 kWh) in the home (HMSO, 1990). Figures in brackets shows annual appliance consumption in a typical home

Space Heating	19.9
Refrigeration	17.0 (302 kWh)
Water Heating	12.8
Cooking	8.3 (660 kWh)
Lighting	8.0 (358 kWh)
Television	5.1 (234 kWh)
Kettles	4.8 (247 kWh)
Washing Machines	3.8 (197 kWh)
Miscellaneous	9.9
Total	89.6

the 100 million lamps sold each year in the UK 98% are incandescent bulbs and over 90% of the energy consumed by these bulbs is lost as heat dissipation. A bus-controlled system with sensors responding to the occupancy of a room could be designed automatically to dim lights when an empty room is detected.

Approximately 36% of the total electricity supplied in the UK (a total of 90×10^9 kWh) is used in the home. A similar proportion is used in the USA. Some form of automatic metering of consumption will be required and, in addition to energy efficient design of domestic products, serial bus control of appliance will enable operating schedules to be implemented to take advantage of electricity pricing schemes.

Spot pricing (Tester, Wood and Ferrari, 1991; Wacks, 1991; Surrat, 1991) of electricity for control of energy-consuming appliances and storage devices is receiving increasing attention. The spot price is the price at which the utilities buy and sell power through the electricity supply grid. The higher the demand for electricity, then the greater the spot price. Also, on days when pollution levels are excessive, if electrical demand rises so high that a utility has to bring on line some generators which are heavy pollution emitters, this could be reflected in a price premium for those generators leading to a higher spot price when they are producing electricity. To respond to the regular updates of the price, the consumer home automation system must be able to interpret the price data supplied by the utility to decide on appliance operating schedules. The control of appliances will involve adjusting the energy consumed by an appliance without necessarily turning it fully off or on. This will ensure that users will always receive some benefit from an appliance. The use of a serial bus to link the home supervisory system with appliances will clearly facilitate the implementation of this demand side management of energy consumption.

However it should be recognized that most of the energy cost reduction is achieved by minimizing the running costs of space heaters, water heaters and refrigerators; and the incentive will be much less to extend the scheme to other appliances which will in many cases be inconvenient for users to accept.

Remote metering using smart meters has been under investigation since the mid-1970s. Dooley (1991) and Dettmer (1992) present brief reviews of this work. Communication signals can be transmitted to user locations by telephone lines, by mains power supply cables or they can be transmitted by a radio link. Low-frequency (100 Hz–2 kHz) signals superimposed on the electricity mains has been successfully used for remote switching (often called ripple control) of street lighting and heating systems. In the UK the BBC's longwave radio transmission has been used to carry control signals enabling the operating times of water and night storage heaters to be staggered to take advantage of cheap-rate electricity. The use of telephone lines is attractive but the need to pay a third party has been perceived to be a significant disadvantage. In recent years low-voltage mains signalling and low-power radio links have been developed to the stage where successful systems have been demonstrated.

Typically a mains signalling system will involve two key components: a customer unit and a controller located in the transformer substation serving a group of houses. Recent research has shown that spread spectrum techniques can be adopted for mains signalling and allows reliable communication over the low-voltage network despite the effect of time-varying and reactive loads on the communication channel. All European systems must operate within the CENELEC-defined, 9–95 kHz, frequency band for utilities and their licensees.

Low-power radio can offer a cheap alternative to mains signalling. It has been used successfully in the USA operating at 900 MHz. In the UK the abandoned 405 line television broadcast frequency (183.5–184.5 MHz) has been adopted and a cellular network of receivers is used, with each cell typically having a 500 m radius.

The Integrated Services Digital Network (ISDN) has been discussed in Section 4.3. ISDN will bring voice, data and video services into the home. (Hatamien and Bowen (1983) describes an early home broadband network). ISDN offers a much higher communication bandwidth than that currently available over the public switched telephone network and its powerful message-oriented signalling method is well suited for service expansion. Lin (1990) has proposed an ISDN-based Home Communication System which is designed to accommodate the ISDN central office-based services, as well as home information products.

Cawkell (1991) commented that it is remarkably easy to describe scenarios for futuristic home systems if no predictions are made about the time of their arrival. 'You can dream-up what you like but if it is not going to happen until,

6.2 DOMESTIC BUILDINGS

say 2080, when almost nobody now will see it, it is not really of much interest or value except to the readers of a science fiction magazine'. Nevertheless the need for an interconnecting bus for home products was soon recognized. For example Phillips pioneered (in 1980) a simple two-wire interconnect and control scheme, the D2B bus, which enabled the remote control of domestic equipment. In addition, in the USA the Electronic Industries Association (EIA) formed the Consumer Electronic Bus (CEbus) Standard Committee in 1984 with the aim of creating a standard for control of home appliances (Douligeris, 1993). Japanese work in this area has closely paralleled the work of the CEbus group. The CEbus and the Japanese Home Bus System are discussed in Section A9. Sufficient interest had been generated by 1985 for a whole issue of the IEEE magazine *Spectrum* to be devoted to home automation (Jurgens and Perry, 1985). By 1993, Sakamura (1993) was able to report that the Japanese TRON Intelligent House Project had progressed to the construction of a pilot house with around 1000 built-in computer elements. It would appear that the time interval of significant technical advances is much shorter than a human lifetime.

In 1987 a consortium of European companies (Thorn-EMI, Phillips, Thompson, Siemens, GEC, Electrolux) started to work on a standard defining the way that signals would be exchanged between appliances connected to a home network. It is interesting to note that these companies accounted for approximately 80% of domestic appliances manufactured in Europe (Tritton, 1988). This group of European companies became the ESPRIT Home Systems Consortium (with the addition of, for example, British Telecom). Their standards work has been promoted by the European Home Systems Association and it has been submitted to CENELEC as a potential European and International Standard. This work is clearly manufacturer-driven. Cawkell (1992) has noted that standards would benefit from a much greater user involvement.

A home automation system uses serial networks and intelligent sensor (and actuator) concepts. It involves the connection of appliances and other domestic devices to the serial network and allows their control via a PC-type interface (e.g. with graphic display type of software), hand-held terminals or spoken instruction. In addition, interaction of devices with other devices and with external organizations via an interface to radio, utility and telephone networks will enable an autonomous mode of operation. The network is likely to use a wide range of interconnect media including, mains supply cables, twisted-pairs, co-axial cable and infra-red and radio frequency wireless links. The metallic pipe circulating hot water to radiators in a house central heating system has also been investigated as a communications path for home automation applications (Pellerin, Brissand and Grange, 1990). To define communication protocols a reduced stack OSI communication model, using layers 1, 2, 3 and 7, is the minimum requirement for the home automation system. A block diagram of

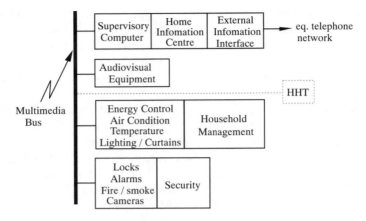

Figure 6.1
Diagram of a home automation system

a comprehensive home automation system is shown in Figure 6.1. The main features of such a system can be summarized as follows:

- a multimedia interconnect capability including twisted pairs, electricity mains cable, coaxial cable, fibre optic links, and infrared and low power radio links
- an interface with the electricity, gas and water supply organizations (plus management systems for the consumption of electricity, water and gas which take advantage of supply price fluctuations)
- a space heating system which provides a programmable zoning facility and may include 'smart' windows (Tester, Wood and Ferrari, 1991) to take advantage of solar energy
- a wider bandwidth interface (e.g. ISDN) to facilitate business, shopping and entertainment interactions with external organizations
- appliances and devices able to connect to a serial bus without requiring reconfiguration of the system (this, of course, will require a communication standard to be established and widely accepted)
- a security system to protect against intruders and hazardous condition, e.g. fires and water ingress
- a user-friendly interface system for house occupiers which will include operation from a PC-style terminal, a hand-held terminal and from speech input.

An evolutionary approach to the development of home automation systems can be followed once a standard communication technique has been established.

6.2 DOMESTIC BUILDINGS

There are many reasons why home automation has been slow to develop. The lack of a widely accepted communication standard is one reason, but high cost (not just of devices but also the costs associated with installing and maintaining a system) and the complexity of the user interface have a significant impact on users (Butler, 1991).

In general, user interfaces for electronic equipment are inadequate. Electronic equipment has become more complex due to increased product functionality. This has arisen from the use of microprocessors coupled with digital controls and panels. The user interface often consists of both a physical and logical component. The physical interface is the means by which the user controls equipment and receives feedback about its status. However, the physical interface may be unable to represent the full functionality of a piece of equipment. In such cases the equipment requires a logical interface which will allow the user to access a larger number of functions. In effect, the physical controls become separated or buffered from product functions. The result of pressing a key may be dependant upon a previous key or the current state of the equipment. The operating and programming of such equipment has become an increasing burden upon the consumer. It is not surprising that potential customers do not understand home automation and are unaware of its potential benefits.

The basic user interface requirements can be summarized as follows.

- The physical interface requires a simple understandable panel with few controls to understand and operate the equipment. There should be no meaningless icons or abbreviations.
- There must be the capability to trap and recover from user errors made with the logical interface. Another useful feature would be to provide the user with meaningful information about the nature of the error.
- The mapping between the physical and logical interfaces must be displayed to the user whilst they are operating the equipment. The user cannot be expected to memorize combinations of key presses or frequently refer to the instruction manual.

Artificial intelligence techniques (Evans, 1991b) could be used to help solve the problem of the user interaction with equipment. Booth (1990) reviews work in the area of human–computer interaction and provides a useful annotated bibliography of the published literature in this area. Such techniques could lead to home automation systems with the ability to recognize verbal commands and even allow them to adapt to the changing requirements of a family within the home. Such features would encourage the use of home automation systems.

6.3 COMMERCIAL BUILDINGS

This section will focus primarily on the office working environment. Factory automation was considered in Section 4.3 and domestic automation was addressed in the preceding section. The intelligent office building is commonly (and advantageously) viewed as a system with the following subsystems:

- computer and telecommunication systems with a recognizable command and control centre
- hazard alarm and security system
- energy management control system (including the conventional heating, ventilation and air conditioning (HVAC) system)
- electrical and communication wiring infrastructure
- electrical power supply system (often including an uninteruptible power supply (UPS))
- a utilities system.

Finley, Karakura and Nbogni, (1991) note that the aim of building automation is to increase worker comfort, productivity, creativity and security; to control energy sources so as to minimize costs and maximize individual comfort; to facilitate overall building management; and permit building owners to realize profits from new sources such as computer and communication services and shared tenant services.

Each subsystem could be very complex and involve a wide range of sensors and actuators. For example a security and emergency control subsystem will have sensors to detect fires and smoke, poisonous gases, and the integrity of windows and fire doors. In addition, for access to the building, a card-controlled system could be used and use may be made of finger print, voice print and retinal scan techniques. Lifts, escalators and automatic doors will require actuators and specialist sensing devices. Whether it consists of twisted pair, coaxial cable or fibre optic cables, the hard wiring for human, computer and sensing communications within a building is expensive and troublesome to install, maintain and (especially) change. Beneath much of today's increasingly dense office electronic environment lies a tangled, confusing, virtually unmanageable maze of wiring. Although most major telecommunication suppliers now offer greatly improved systematic premise wiring systems, building wiring remains a difficult chore and cabling occupies a great deal of space in the modern building (Freeburg, 1991).

Serial communication networks will simplify cabling problems and facilitate the use of information technology and automation techniques in buildings. Fey (1991) describes a network of distributed stand-alone controllers and operator workstations operating over a two-tiered LAN (as shown in Figure 6.2). In

6.3 COMMERCIAL BUILDINGS

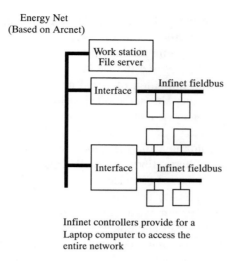

Figure 6.2
A two-tiered LAN topology used in commercial building applications

this network, the backbone bus Energy Net provides the system with a high-speed token-passing communications protocol, based on the Arcnet protocol (see Section A.2). The backbone network is extended by a fieldbus called Infinet. Any of the Infinet controllers can be programmed or interrogated through a laptop computer via a built-in connector located in each controller. Each of these additional users has complete access to the entire network. For large networks with hundreds or even thousands of controllers, service personnel are able to respond to problems from any location in the building by simply connecting to the nearest controller with a portable computer. Fey (1991) reports an application of the Energy Net–Infinet system to a seven acre commercial site, which provides direct digital control of all HVAC equipment, access and lightning control, ramp and traffic control in the parking facilities and complete integration of the smoke and fire control systems for more than 13 000 points with more than 2000 stand-alone controllers. An interesting feature of this system is that it uses a natural language (using easily understood English statements) building control language.

Serial network standards for commercial building applications are still under development. The IEC fieldbus standard will no doubt find applications in this area. Manufacturers are adopting a multiprotocol approach. For example Clapp and Churches (1993) describe the Satchnet system which is a proprietary communications network. The protocol of this network is freely available, thus allowing third party companies to integrate Satchwell products within their own system. Satchsoft, the software that runs on the control terminal of the network, has a repertoire of six different protocols; three are proprietary to Satchwell, two are proprietary to competing companies, and the last JBUS, is an open protocol that is used in France and the USA.

Given the expense and limitation of wired networks, the use of radio technology becomes an obvious alternative. The cellular technique has particular merit because the working environment can be partitioned into microcells (each with an allocated frequency) defined by floors and major walls. UHF spread spectrum in-building communication methods suffer from critical bandwidths and spectrum reuse problems. Fortunately 18 GHz microcellular in-building communication networks have been made possible by developments in CMOS and Gallium Arsenide (GaAs) semiconductor technology (Freeburg, 1991). Also the antenna for the 16 mm wavelength of an 18 GHz transmission can be designed to be relatively small and compact. Microwave transmissions diffuse thoroughly through a network microcell (with dimensions of about 30 m) using only a minimum of radiated power and remains confined within the microcell so that the same frequencies can be reused by another system. In the USA the FCC has allocated the bands 18.820–18.870 GHz and 19.160–19.210 GHz which is enough for ten 10 MHz channels. Each channel will be able to offer high-bit-rate two-way digital transmission using the time division multiplexing technique.

A human-centred approach to automation was briefly discussed in Section 3.2. It is interesting to note that similar reservations about the high-level application of automation methods have arisen in the development of the intelligent building concept. Powell (1988) feels that intelligent building systems must be dynamically responsive, and provide a co-ordinated strategy in which users can actively control their own environment and their access to information in the system (not simply that predefined by the system manager). As discussed in the preceding section the design of the user interface to the home system is of critical importance to its success. This is clearly true also of the commercial building but in this case two classes of user can be simultaneously active: building management will want to control the building to achieve some form of optimum financial performance, and users working in the intelligent building will appreciate (and benefit from) access to the system that will provide them control of their immediate environment. This implies a high level of flexibility in the building control strategies with an ability to gracefully resolve conflicts that will inevitably arise between the two classes of user. It also implies greater access to the building communication network to enable users in predefined zones to be able to exercise control over their immediate environment. Clearly both classes of user commands will have to be defined, and considerable advantage will be derived from adding a user layer (8th layer) to the communication model for the intelligent building serial communication network.

Atkin (1988) discusses examples of intelligent buildings and notes that the major activity is in Japan. Notable work in the UK includes the Royal Bank of Scotland building (Islington, London) and the Lloyd's of London building in the City of London. In Europe the development of intelligent buildings is constrained by the enormous influence of industrial democracy on the quality of working life. This has the effect of requiring intelligent buildings to be designed to allow each individual worker to adjust lighting, air conditioning and access data.

Japan's plans for the $400 billion, 25 000 acre Tokyo Bay development will no doubt further enhance their technological capability. In Japan there is considerable interest in building intelligence, particularly from their Nippon Telegraph and Telephone (NTT) Corporation. Japan shares similar market conditions with North America (e.g. deregulation of NTT in 1986). However their vision is of interactive communication networks rather than a collection of semi-autonomous building units. Also, behind the obvious commercial motives there is an industrial culture that is moving towards a complete electronic capability.

In North America, Shared Tenant Services are emphasized. Of interest here is the value added by information technology services to multitenanted property. The deregulation of the US telephone monopoly provided a large variety of choice for users of telecommunication services. Each large building can, in effect, become a private utility company as far as telecommunications are concerned, run for the benefit of tenants and the profit of the landlord.

Architects are continually exploring the application of modern engineering technology. Groak (1992) discusses the intelligent building and introduces the intelligent site as a logical extension of the current automation activity in buildings. Richard Rogers (1991) talks about buildings interacting dynamically with climate and goes on to say 'more like robots than temples, these apparitions with chameleon-like surfaces insist that we rethink yet again the art of building. Architecture will no longer be a question of mass and volume but of light weight structures whose superimposed transparent layers will create form so that constructions will become dematerialised'. In a later passage he notes that 'in the case of architectural structures, responsive systems, acting much like muscle flexing in a body will reduce mass to a minimum by shifting load and forces with the aid of an electronic nervous system which will sense environmental changes and register individual needs'. These futuristic constructions will clearly make extensive use of serial networked instrumentation.

6.4 STANDARDS

A large number of standards groups have (and continue) to work to develop communication standards for home automation applications. In the USA the

National Association of Home Builders proposed, in their Smart House Project, a closed-loop wiring system to carry power, in-house communication, and control signals to every outlet, light fixture and appliance in a house. Smart House systems are designed specifically for new houses. Again in the USA the Consumer Electronic Group of the EIA have published their Consumer Electronic Bus (CEbus) standard (see Section A.9). In Japan, organizations again within the EIA, have developed the Home Bus Systems (HBS) standard (this is also discussed further in Section A.9). Both the CEbus and HBS standards were intended for single-family homes. The Japanese group have also defined a Super Home Bus System (S-HBS) that is used to interconnect apartments within a building (Hamabe *et al.*, 1988a, b). CENELEC is co-ordinating the standardization of the European Integrated Home System (IHS). This work is based on a European Community ESPRIT research programme which was originally instigated by white goods manufacturers (Baxtor, 1988; Coschieri and Troian, 1989). Other significant European standardization activity includes:

- European Installation Bus: this was originated by Siemens for building management; it is promoted by the Electrical Installation Bus Association and it has been submitted to CENELEC
- Batibus: this was originated by Merlin Gerin for building management; it is similar to EIB but it is intended for smaller lower-cost networks, and it has a low bit rate (4.8 kb/s rather than 9.6 kb/s)
- D2B: this was originated by Phillips for audiovisual control, primarily within single rooms.

The overall serial networks activity in the home automation area is clearly considerable. A widely accepted international standard is required to spur the development of home automation products.

A wide variety of serial networks are used in intelligent commercial buildings. Arcnet (see Section A.2) and other proprietary networks are used. The IEC Fieldbus standard was originally specified for applications including intelligent buildings, but the intended applications now appear to be factory automation and process control.

Proprietary serial networks continue to be developed. An example of a recent network is the Echelon local operating network (LON). The Echelon Corporation Inc. (California, USA) was established to provide a complete off-the-shelf solution to design and implement serial control networks. The Echelon Corporation aims to establish its LONWORKs system as the *de facto* standard for all forms of control systems, including intelligent buildings of all types. The local operating network is discussed in Section A.10 and Appendix C.3.

7
Transport

7.1 INTRODUCTION

Serial bus application in transport systems are characterized by requirements for high-speed operation and high reliability. Bit rates are commonly 1 Mb/s although operating restrictions often lead to an effective rate that is a quarter of this value. This chapter discusses the extensively used military avionics bus defined by MIL-STD-1553B, the ARINC 629 bus used in civil aviation applications and the various bus systems used by commercial car manufacturers. The extreme competitive pressures of the car market are no doubt responsible for the large number of serial bus systems designed for in-car applications. In this chapter three of these bus systems will be discussed, namely: the controller area network (CAN), the vehicle area network (VAN) and the SAE J1850 bus. The stringent performance requirements for in-vehicle networks has led to the CAN being adopted for a range of general industrial applications. This chapter concludes by briefly reviewing the use of serial communication techniques in road automation.

7.2 MILITARY SYSTEMS

Work to define a serial bus standard for electrical cables used in military avionic applications started in 1968, and this led to MIL-STD-1553A in 1975 and its updated version MIL-STD-1553B in 1978. Equipment based on these standards has proved to be flexible and very reliable. MIL-STD-1553B has been adopted for use in all types of military equipment including avionics, ships, tanks and some missiles. It is used in military systems worldwide, except in the former communist block countries. The standard is managed by the Society of Automobile Engineers, Avionics Systems Division.

A fibre optic version of MIL-STD-1553B was published in 1987 and later became MIL-STD-1773. This was a natural progression since the reduced weight and EMI immunity of fibre is an attractive feature in many applications. A shielded 1553B bus can offer very high levels of noise immunity but this is achieved at the expense of increased weight. The 1773 standard was designed to enable a fibre optic system to replace a 1553 electrical system with no impact on existing equipment other than requiring the replacement of electrical transmitters and receivers with optical devices.

A 1553 multidrop bus uses relatively short Manchester encoded messages transmitted over a twisted pair. The twisted pair is specified by the standard to be shielded and devices are transformer-coupled with isolation resistors used to prevent short circuit failures. A bus controller manages the information flow to remote terminals that interface one (or more) field-located subsystems (or devices) to the bus and respond to commands issued by the controller. The widely recognized communication security provided by this bus is largely due to the command/response protocol used to control access to the bus. A typical communication sequence would start with the bus controller sending a command to a remote terminal. The addressed remote terminal then either accepts data or transmits data as commanded, and then responds with a status word indicating that all of the words were received and checked for parity. It should be noted that data transmission checking at bit, word and message level with two different command and data synchronization headers guarantees a very low residual error rate. The status word confirms to the bus controller that a communication link has been reliably established. A 12 μs time-out period is set by the bus controller for the remote terminal to respond. More information about the 1553 standard is given in Appendix A3.

Although the 1553 bus operates with a bit rate of 1 Mb/s, response and transit time limitations introduced by the command/response protocol limits the effective bit rate to a much lower value. Typically, throughput will be about 250 kb/s with a burst mode maximum of about 800 kb/s. If greater speed is required then extra bus systems can be added to increase throughput.

For normal implementations of the 1553 bus in avionic applications the majority of data are transferred cyclically. A major cycle of typically one second is used in which all data are transferred at least once. This period is divided into a number of minor cycles, usually a power of two and typically 64, giving a minor cycle rate of 64 Hz. Figure 7.1 shows major and minor cycles in a typical 1553 bus access sequence.

Many bus systems will use a fixed schedule of data transfers. The detailed design of the schedule is based on the largest and smallest iteration and allowable latencies. The slowest iteration rate (the least common multiple of the faster iteration rate) is normally defined as the major cycle. The minor cycle is normally the frequency of the most rapidly transmitted periodic data. Minor cycles can be defined by binary or decimal division of the major frame. If the major frame is 1 s long and there are 64 minor cycles, then each minor cycle is $\frac{1}{64}$ (15.625) ms long. If a transaction is required to occur eight times per

7.2 MILITARY SYSTEMS

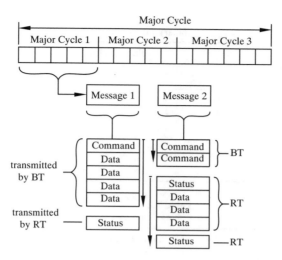

Figure 7.1
Major and minor cycles in a typical 1553 bus access sequence

second, it must occur during one of the first eight minor cycles ($\frac{64}{8} = 8$) and every eight minor cycles thereafter (e.g. 3, 11, 19, 27...).

Sophisticated polling protocols are used in practice. For example, there is the binary interval method. This allows remote terminals to be polled at different intervals while retaining unoccupied polling slots for further expansion. There is always a further vacant slot in which to insert further polls; however these occur at ever increasing intervals. Table 7.1 demonstrates this method. There is also the rolling piano key method. This allocates different polling frequencies to different functions within a subsystem. To smooth the loading on the bus, time frames which contain different rate data are staggered for successive remote terminals. Table 7.2 demonstrates this method.

Provision must be made for aperiodic message transactions. If the aperiodic message originates in the bus controller than it can initiate the necessary transaction immediately if the bus is idle, or it can insert a transmit command into the transmission list so that the message will be transmitted during the next available time slot which is not dedicated to a periodic transmission. If an aperiodic message originates in a remote terminal then the remote terminal interrupts the bus controller by setting the service request bit in the status response associated with a periodic transmission. This is recognized by the bus controller and it then uses the first available idle period to take appropriate action.

Scheduling serial network resources to satisfy so-called hard real-time requirements is a concern in all the application areas discussed in this book. The 1553B

Table 7.1
Binary interval method

	Poll Cycle															
RT	0	1	2	3	4	5	6	7	8	9	10	11	12	13	14	15
1	+		+		+		+		+		+		+		+	
2	+		+		+		+		+		+		+		+	
3	+		+		+		+		+		+		+		+	
4		+				+				+				+		
5				+								+				
6								+								

Table 7.2
Rolling piano key method

		Time Frame															
Rate	RT	0	1	2	3	4	5	6	7	8	9	10	11	12	13	14	15
50	1	+	+	+	+	+	+	+	+	+	+	+	+	+	+	+	+
	2	*	*	*	*	*	*	*	*	*	*	*	*	*	*	*	*
	3	#	#	#	#	#	#	#	#	#	#	#	#	#	#	#	#
25	1	+		+		+		+		+		+		+		+	
	2		*		*		*		*		*		*		*		*
	3	#		#		#		#		#		#		#		#	
12.5	1	+				+				+				+			
	2		*				*				*				*		
	3			#				#				#				#	
6.25	1	+								+							
	2		*								*						
	3			#								#					
3.12	1	+															
	2		*														
	3			#													
1.56	1	+															
	2		*														
	3				#												

avionics-related work was an early contribution to this area. Liu and Leyland (1973) discuss scheduling algorithms, and more recently Cadeira, Siebert and Thomesse, (1993) presented a FIP-oriented discussion of the scheduling of both tasks and network traffic in a fieldbus-based real-time system.

A detailed discussion of the current status of the 1553 standard documentation (including fibre optic advances) and available hardware is presented in the Proceedings of the Conference on MIL-STD-1553B and The

Next Generation, held in London in November 1989. (This is published as ERA Report No. 89-0591 by ERA Technology Ltd, Cleeve Road, Leatherhead, Surrey, KT22 7SA, UK). Papers are included on non-military applications such as: control of the large European accelerators (Rausch, 1989), a 1553B industrial fieldbus (Stone, 1989) and a hybrid twisted pair/optical fieldbus based on MIL-STD-1553B (Lytollis, Jordan and Kelly, 1989). The Edinburgh University low-power, twisted pair/fibre, optic fieldbus was based on MIL-STD-1553B (Jordan, Lytollis and Kent, 1992).

7.3 CIVIL AVIATION

ARINC 629 defines a digital wire or fibre optic bus which allows avionic subsystems to transmit and receive digital data using a standard communication protocol. The ARINC specification 429 (Mark 33 Digital Information Transfer System), adopted in 1977, has performed satisfactorily and with high reliability. The objective of ARINC 629 was to reduce airplane wiring and equipment interfaces, and simplify the implementation of centralized maintenance equipment where large and rapid data exchanges are required.

One of three media types may be selected for a given application: namely unshielded or shielded twisted pair (current mode), shielded twisted pair (voltage mode) and fibre optic. With the current mode bus wires pass through a coupling transformer unbroken (i.e. no connectors are required) and a clip-on coupling arrangement is possible (this technique was subsequently adopted by the Norwegian ISIbus group). A maximum of 120 terminals are specified on a linear bus (maximum length 100 m) with T-stubs of length up to 15 m.

Carrier sense multiple access/collision avoidance (CSMA/CA) is used with Manchester II bi-phase coding. The transmission bit rate on the wire bus should be 2 MHz. A dedicated bus controller is not required and bus access control is distributed among all of the participating terminals. A basic protocol allows terminals to communicate at constant intervals (periodic mode) during normal operation and, if the bus is overloaded, it will automatically switch to an aperiodic mode. A combined mode protocol provides for priority access for aperiodic data. The combined mode automatically limits the throughput of aperiodic data in a progressive fashion (from the lowest level upwards) according to the available time without affecting the periodicity of normally periodic data transmission. The ARINC 629 standard is further discussed in Section A.5.

7.4 VEHICLES

The number and complexity of electronic control units (ECUs) in cars has steadily increased over the past two decades, leading to a change in the wiring architecture used to connect the whole system together, from conventional wiring architectures using a considerable amount of wire to a modern architecture using a serial data highway compact. Industry experts are predicting that 25% of vehicles will be using serial data highways by 1995. The serial bus must be designed to satisfy the data rate required by sensors and control systems, and the latency introduced by using the bus must be held reliably to a low value. The sample period required by signals in the car system and the maximum time that the message may be delayed (latency time) before its instruction or data could cause the system to perform badly are important design parameters. Fast signals such as engine speed, fuel rate and braking torque require sampling periods and latency times of about 10 ms. Slower signals such as coolant and air temperature and gear change requests require sampling periods in the range 100–200 ms and latency times of the order of 35 ms.

The Society of Automotive Engineers (SAE) defines three categories of data transmission for use in a vehicle, namely:

- a low-speed option (< 10 kbit/s), aimed at general multiplexing of body functions — Class A
- a medium-speed option (10–100 kbit/s), aimed at sensors which need to share data — Class B
- a high-speed option (100 kbit/s–1 Mbit/s), aimed at real-time control data, such as engine management or ABS braking systems — Class C.

An ISO classification has only two categories, namely: low speed (< 125 kbit/s) used for body control systems; and high speed (> 125 kbit/s) used for real-time control.

Any vehicle which uses a serial data highway should be able to use equipment from many different manufacturers simultaneously without compatibility problems. This leads to a requirement for a large number of ECUs which simply replace existing ones in the way that car radios can be swapped. An identifier block, indicating which node is to receive the message, is used to allocate each ECU model a unique number which it uses in network communications. The serial bus systems discussed in this section all use message identifiers and often more than 10 bits are used. A large number of identifier codes prepares for a large number of possible market products.

Another common feature of the serial bus systems described in this section is the use of a technique called bit-wise arbitration to prevent collisions while still allowing the highest-priority message access to the bus. Here, the transmitting node monitors the bus and compares the sent bits to the actual bits on the bus. If the node detects a difference, then it stops transmitting and enters the receive mode, leaving the higher priority message undamaged. (This should be compared with a conventional LAN collision detection scheme where all of

7.4 VEHICLES

the nodes stop transmission after a message collision is detected.) The transmitter sending the lower-priority message may re-transmit its message once the bus is clear. Address priorities are arranged to ensure that bit-wise arbitration favours lower numbers over higher ones, allowing the most important signals (usually those with low latency times) access to the bus in a case of contention. It is claimed that bit-wise arbitration is a more reliable method to achieve a deterministic latency time with minimum overhead for higher priority messages.

The European controller area network (CAN), the French vehicle area network (VAN) and the American network defined by SAE J1850 have all been developed specifically for in-vehicle networks. Each network offers the usual benefits (e.g. reduced wiring complexity) but there are marked differences in their data transfer formats.

VAN is a French project developed for vehicles by Renault and PSA (Peugeot and Citroen), and adopted by the Bureau de Normalisation Automobile (BNA) in June 1989. It is being processed for publication by the Association Française de Normalisation (AFNOR) and it has been proposed to the ISO. Renault intend to use the VAN protocol in all their production vehicles by 1994. The ISO have classed this standard as low speed (< 125 kbit/s), even though it is claimed to have a data transfer rate of 250 kbit/s. Of the three major standards, VAN is currently the only one which can be used for SAE classes A, B, and possibly C, simultaneously. Integrated circuits are being developed for the VAN system.

VAN has been designed to be able to perform logical addressing, point to point acknowledgement and master/slave operations. As an example of VAN's general multiplexing and master/slave capabilities, the ASIC VAN developments have created a remote-controlled switch (RCS) which has the ability to receive and transmit a byte using the VAN protocol. Other nodes and ECUs may write to this switch which, in turn, controls up to eight intelligent power switches. The RCS does not contain any complex or costly microprocessors, just the VAN core and some I/O capability. By using systems such as RCS, the overall cost and complexity of the vehicles electrical control units will be greatly reduced.

A VAN data frame is transmitted using a variant of Manchester bit encoding, and it consists of nine data fields in the following order:

SOF, IDEN, COM, DATA, FCS, EOD, ACK, EOF, IFS

SOF is the start of frame field. It synchronizes the receiver clocks over the whole system to make sure that all the receiving nodes are aware of the message. It also provides time-reference generation, allowing cheaper modules and ECUs to be manufactured. A SOF field cannot be sent if another node is already using the bus.

IDEN is a 12 bit identifier that specifies which node the message is intended for. The 12 bits of IDEN mean that there may be up to 4096 nodes on the network.

COM contains command information associated with the frame, describing its contents and function. It also determines whether the receiving node should reply within the frame or not.

DATA is an optional field which has arbitration capability, containing up to 28 bytes of data. The arbitration capability will be used if the data field is to be a reply from a number of other nodes. In such a case, a priority identifier similar to IDEN will be sent before the data.

FCS is a field check sequence of 15 bits. It uses cyclic redundancy checking (CRC) to check the data. The check ensures that the IDEN, COM, DATA and FCS fields have been received correctly.

EOD is an end of data field.

ACK is an optional acknowledge field, sent by the receiving node to indicate that it has received the message correctly. If the message has an error, this field will indicate that the transmitting node should send the message again.

EOF is the end of frame field.

IFS is an inter-frame space.

If a transmission request is included in the COM field, the DATA, FCS, EOD and EOF fields are generated by the node which the message is addressed to. This procedure is known as in-frame response.

The J1850 standard was designed by the Society of Automotive Engineers (SAE) to cater for a class B network. A J1850 data frame can be transmitted at either 10.4 kbit/s using variable pulse width (VPW) encoding and a single ground referenced wire or at 41.6 kbit/s using pulse width modulation (PWM) and a dual wire differential voltage. VPW is an encoding scheme developed by General Motors. It has several advantages over ordinary pulse width modulation, one of which is that VPW generates half as many EMI (electromagnetic interference) producing voltage changes for a PWM bus operating at the same speed. VPW can achieve one state transition per bit by the definition of four bus characteristics: an active or passive state and a long or short pulse.

The J1850 data frame is made up of seven data fields in the following order:

SOM, DATA, ERR, EOD, RSP, EOM, IMS

SOM is the start of frame used to synchronize the receiving nodes.

DATA has 101 bits of data with optional error detection data. Its possible data content could be a message or address identifier, 8 bit data values, 7 bit data values with 1 bit parity or a form of error checking data.

ERR is an 8 bit cyclic redundancy check (CRC).

EOD is an end of data field.

7.4 VEHICLES

RSP is the response field, and, if used, it begins immediately after the EOD field. There are three possible forms that the response bytes may take. They can consist of nothing, an 8 bit acknowledgement identifier (with arbitration) or one or more response data bytes followed by an ERR byte.

EOM is the end of message signal.

IMS is the inter-message separation, used to give the various nodes a chance to synchronize to each other during back-to-back message operation. Before any node may transmit a SOM field, either EOM and IMS have expired or EOM has expired and another rising edge has been detected.

The bytes of the message can be separated by IBS (inter-byte separation) to allow for unbuffered interface hardware or to alleviate critical timing constraints. With IBS all data and response bytes must be followed by zero to a number of dominant bits determined by the specification, followed by a very brief period (less than a bit) of a recessive bit to allow the nodes to resynchronize. The dominant bits will allow bit arbitration to continue into the next byte. When the interface hardware contains a multiple byte message buffer, all the data bytes and the response bytes may be transmitted one after the other without interruption. For systems that use IBS, all the receiving nodes should be able to receive a message that does not transmit data bytes and response bytes without interruption.

CAN (controller area network) was developed by Bosch (in Germany) and is now approaching the status of a *de facto* standard for car manufacturers to use, having been adopted by motor manufacturers such as Rolls Royce, Mercedes Benz and the Rover Group. CAN's flexibility and high performance has led to some non-automotive uses, such as weaving machines in the textile industry, elevators and security systems. Some of the major component manufacturers (Motorola, NEC, Intel and Phillips) have introduced product ranges which use the CAN standard and are claimed to be compatible with one another on the same bus. CAN is capable of operating at speeds of 15 kbit/s to 1 Mbit/s (SAE classes B and C) and, as a result, it is considered by many to be the only realistic option for real-time control. Two versions of CAN, Basic CAN and Full CAN have been developed.

With Basic CAN, the CPU performs the message handling with only a little support from the CAN controller (the controller only handles the message filtering). The communication data is represented to the CPU by a receive/transmit buffer. This level can handle any number of messages, but it places a high burden on the CPU's processing time.

Of the three networks discussed here, CAN offers the best performance in error recovery, transport capacity, and noise immunity. Its high-speed capabilities and real-time control of information sharing between ECU's has led to it becoming widely accepted by various automotive manufacturers and their suppliers. CAN controllers are commercially available from many manufacturers. CAN is discussed in more detail in Section A.12.

In the US there are three standards. They are nearly identical to the J1850 standard and are being developed separately by General Motors, Ford and Chrysler. They are not compatible with each other or with J1850.

In Japan, Mitsubishi has developed its own multiplexing scheme called MICS (Mitsubishi Intelligent Cockpit System) which is based on the J1850 standard. Mazda has also based its multiplexing scheme, called Palnet (Protocol of Automotive Local Area Network) on the J1850 standard. Toyota is working on a multiplexing system which is claimed to be an improvement on CAN, but it uses a similar method of bit transmission (1 Mbit/s with NRZ encoding and bit stuffing).

In Europe, the only automotive manufacturer that has not opted for either VAN or CAN is Volkswagen. They have been working on a proprietary multiplexing system.

7.5 ROAD AUTOMATION

It is difficult to imagine a situation in which vehicle usage in the major urban areas will reduce sufficiently to allow traffic flow to be managed with a low-level automation system. The rule appears to be that as population density increases, traffic density increases to just below the density which causes major hold-ups and totally unacceptable journey times. It is not surprising, therefore, that a very large world-wide activity has developed to investigate what has been called the intelligent vehicle-highway system. An overall systems engineering approach is required that will lead to the use of smart cars on smart highways. The serial highways discussed in the previous section will of course facilitate this construction of intelligent systems in the car. Interaction with the intelligent highway will require serial links to be established between the serial network in the car and the highway management system.

A number of research and development programmes have been established to investigate intelligent vehicle-highway systems (IVHS). Jurgens (1991) reviews the IVHS activity and includes a discussion of the work of the groups Mobility 2000 and IVHS America in the USA. The European activity in this area is reviewed by Catling and McQueen (1991). The European Commission supports two major initiatives: DRIVE (Dedicated Road Infrastructure for Vehicle Safety in Europe) and PROMETHEUS (Programme for a European Traffic with Highest Efficiency and Unprecedented Safety). Clearly in-vehicle systems are entering a period of rapid change with driver status monitoring and vision enhancement, vehicle collision avoidance, intelligent cruise control and dynamic route guidance at an advanced stage in their development. Automatic

7.5 ROAD AUTOMATION

debiting systems are already in use in Norway and the USA; and autonomous electronic display equipment is available. Serial communication networks based on a widely accepted standard (Fisher and Sullivan, 1993) is a critical enabling technology for this work. As for the other application areas discussed in this book the user (driver, maintainer and manufacturer) interface must be carefully designed. This is discussed by Parkes (1992) and he notes that the limiting factor in what can be successfully introduced to the vehicle is not cost, nor dashboard equipment, but the interests and capabilities of drivers.

A wide range of techniques have been investigated to communicate with the car. These include the use of buried inductive loops, modulated infrared beams and additional inaudible modulation of FM radio broadcasts. It is expected that satellite-based (for example, using low earth orbit satellite) communication systems will be increasingly used in vehicle location and navigation.

The inductive loops used to detect vehicles can also be used for communication purposes (Carter *et al.*, 1993). A buried inductive loop becomes effectively the secondary of a transformer with the primary mounted on vehicles travelling over the loop. A double frequency system operating at 50 and 100 kHz is used to transmit from the loop and transmit from the vehicle. Applications include access control with electronic key operated barriers, road intersection control and the management of public transport systems. However inductive loops are expensive to install and tend to have a short lifetime. In urban locations the lifetime of a loop can be two years, whereas for motorway applications lifetimes can be in the range 10-20 years.

An attractive alternative to inductive loops is to use infrared beacons which would typically be mounted on traffic lights. Infrared transmitters are arranged to transmit the same message simultaneously and therefore transmit identical information to all vehicles approaching the intersection. The vehicles themselves can be designed to collect traffic data. For example on-board computers can be programmed to measure journey times for each section of a route and the waiting times spent at traffic signals. This information can be transmitted to the central road management system when a convenient beacon is passed. von Tomkewitsch (1991) describes an infrared beacon system that uses messages defined by the high-level data link control (HDLC) block format. This system has been extensively tested in Berlin. It will be possible to transmit up to 65 kbytes of data to vehicles at a data transmission rate of 500 kb/s.

Navigation and route guidance information can be obtained via FM radio broadcasts. The Radio Data System (RDS) is a system for transmitting data by means of a silent data system, superimposed on normal FM radio broadcasts. The data rate used is further subdivided into four blocks of 26 bits, with each block containing 16 data bits, and one error-correction bit. Messages can be specific instructions to avoid a hazard or general comments about weather conditions and overall traffic management.

It is well known that vehicles travelling in convoys are often involved in accidents. Shladover *et al.* (1991) discuss a spacing control system for use in such situations. They propose the use of an infrared serial link between the rear of one car and the front of the following car. A serial link through the following car enables a communication chain to be formed linking the first vehicle to the last vehicle in the convoy. Control of a vehicle in the convoy requires access to speed and acceleration of the vehicle, distance to the preceding vehicle, acceleration and speed of the first vehicle in the convoy. Each vehicle receives a signal from the preceding vehicle which it uses to calculate a safe separation distance (and speed and acceleration). Received information is passed on to the next vehicle with information arising from the above calculations. Shladover *et al.* (1991) show that a bit rate of 230 kb/s is required to control a 15 vehicle convoy. For safety reasons it is likely that a separate in-vehicle bus dedicated to convoy control will be required. A guaranteed speed of response is desirable so a regular sampling polling method should be used to control access to the transmission media. In practice, the required protocol is rather complicated because the number of vehicles in an convoy changes frequently. Another source of difficulty is the potential interference from neighbouring vehicles that are not part of the same convoy. Shladover *et al.* (1991) proposes to overcome these difficulties by using a code-hopping strategy which would use some information about the location of these vehicles to select the code to be used.

8
Electronic Systems

8.1 INTRODUCTION

Serial techniques are widely used in electronic systems. For example general arithmetic and signal processing techniques have been developed and based on bit-serial techniques (Denyer and Renshaw, 1985). This book is primarily concerned with the use of these techniques to interconnect field-located devices. However serial techniques can be usefully extended down to the smaller scale of printed circuit boards (PCBs) and integrated electronic circuits. Field devices based on the use of internal serial electronic circuits are generally smaller because serial integrated circuits have less pins (and therefore have a smaller footprint) and PCBs are smaller because parallel bus connections are minimized.

This chapter first discusses the transputer, a microcomputer specifically designed for serial operation, and electronic system and sensor applications are then described. The next section discusses the two-wire Inter-Integrated Circuit (I^2C) bus and the four-wire bus systems.

8.2 THE TRANSPUTER

A transputer (Inmos, 1988; Mitchell *et al.*, 1990) is a single chip microcomputer with its own local memory. It has serial links to connect one transputer to other transputers. Figure 8.1 shows a block diagram of the main components of a transputer. Transputers can be used in a single processor system or in networks to build high-performance concurrent systems. The programming language OCCAM and the transputer were developed from parallel design activities. OCCAM provides a framework to design concurrent systems using transputers, its use will maximize the benefits of the transputer architecture and will facilitate the use of the special features of the transputers.

Each serial connection comprises two unidirectional links as shown in Figure 8.2. The use of two wires allows handshake operation with one wire sending data and the other wire used for an acknowledgement signal back from

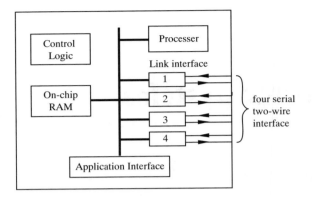

Figure 8.1
Block diagram showing the main components of a transputer

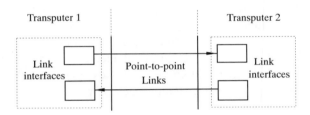

Figure 8.2
Serial communicating link composed of two unidirectional links

the receiving transputer. Point-to-point links simplify communication protocols by eliminating contention problems. Send and acknowledge message formats are shown in Figure 8.3. After transmitting a data byte, the sender waits until an acknowledge signal is received. Some devices allow an acknowledgement to be transmitted as soon as reception of a data byte starts and therefore transmission may be continuous. Communication can be achieved between independently clocked systems, provided that the clock frequency is the same. Adaptor circuits are available to interface transputer links to non-transputer devices (e.g. converting bi-directional serial link data into parallel data streams). All transputer devices support a standard communication frequency of 10 Mb/s and most also support bit rates of 20 Mb/s. Connections (< 30 cm, e.g. on a backplane or PCB) can be established without difficulty. Over longer distances twisted pair connections are used but transmission line effects must be dealt with. If differential mode operations is required RS 422 drivers can be used. Fibre optic links have also been used to link transputer systems.

8.2 THE TRANSPUTER

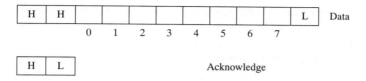

Figure 8.3
Send and acknowledge message formats

A related device in the transputer family provides a transparent programmable link switch designed to function as a cross-bar switch between 32 link inputs and 32 link outputs. A block diagram of this device is shown in Figure 8.4. These devices can be used in a variety of architectures. For example larger switches can be constructed (three 32 way switch devices are required to produce a 48 way link switch) and a fully connected network of 32 transputers can be constructed by using four 32 way link switches.

A transputer-based sensor with local intelligence forms the basis for networks of distributed sensors together with an integrated path for combining the information from different sensors. For control purposes a distributed architecture overcomes the bottlenecks associated with a centralized controller. A serial-to-parallel (parallel-to-serial) convertor enables links to parallel networks and other serial networks to be easily established.

Decentralized and distributed multisensor systems can be established, with each of the sensing nodes maintaining a local model that is most appropriate for the dynamics of its observations. The flexibility of the transputer link scheme allows a node to communicate directly with nodes with observations

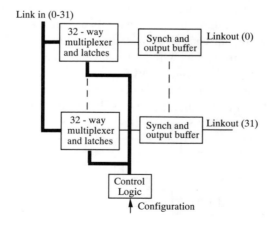

Figure 8.4
32 way link switch

relevant to its own model. Redundant sensor combinations can be connected by a transputer network. In large well-connected arrays the interconnection of nodes can be considered to be arbitrary and dynamically reconfigurable. Hence communication link failures can be overcome by re-routing.

Interest continues to grow in the synergistic use of multiple sensors to increase the capability of intelligent machines and systems (Luo and Kay, 1989). A common requirement of systems using multisensor integration and fusion is that the systems interact with and operate in an unstructured environment without the complete control of a human operator. Integration implies a combined (co-operative) use of information to improve overall performance in some way, whereas fusion indicates that several signals have been combined to form one signal exhibiting some form of improved characteristic (e.g less noise, lower variance). Human operators automatically provide a flexible body of knowledge and the ability to integrate information of different modality obtained through their sensors. Fault-tolerant sensor networks can be constructed based on integration and fusion algorithms and using network reconfiguration algorithms (Iyengar, Jayasimha and Nadig, 1994; Marzullo, 1990; Jordan, Gater and Mackie, 1989; Gater, 1987).

Daniel and Sharkey (1990) describe a 'virtual bus' architecture (Figure 8.5) for robot systems that allows the interconnection of transputer-implemented control processes on the basis of bandwidth, rather than the complexity of a single process. This architecture also facilitates fast error-controlled reconfiguration. Their main aim was to reduce the latency within the robot arm control loop. This led to the distribution of fundamental components of the robot controller between a number of separate processes (e.g. the functions velocity feedback, position feedback, joint strain, arm kinematics, outer-loop force control and dynamic force control would each have a processor). The desirable low-latency feature of the virtual bus architecture was demonstrated as being essential for the control of wide bandwidth loops. This type of fast special-purpose bus will interact with higher-level automation functions directly via a fieldbus or indirectly through a PLC.

The high speed and serial mode of operation of the transputer indicates that it will be useful for implementing interfaces (bridges) between serial networks.

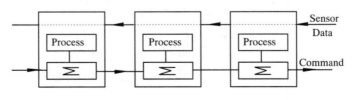

Figure 8.5
Virtual bus architecture

8.3 INTER-INTEGRATED CIRCUIT (I^2C) BUS

Robertson et al. (1990) describe a distributed switch for the connection of PC cards via a high-speed backbone network, with Ethernet bridging as the first application. The backbone network is a fibre optic 75 Mb/s fast ring; so bridged LANs may be physically separated by considerable distances. A transputer is used to control directly the PC LAN via the PC bus.

8.3 INTER-INTEGRATED CIRCUIT (I^2C) BUS

The use of serial interface techniques to interconnect integrated circuits simplifies a printed circuit board layout since the number of I/O pins is reduced. With fewer digital lines generating noise, digital feedthrough to analogue circuitry is significantly reduced. In addition it is easier to route the reduced number of I/O lines away from sensitive analogue circuits. Serial digital interfaces are easier to isolate with optical or transformer techniques, which cost less compared with either parallel digital or analogue isolation. This type of isolation is important for the successful operation of field instrumentation, since it provides a convenient method of accommodating different ground potentials, breaking ground loops to prevent noise and preventing contact with high voltage levels.

The I^2C bus was designed by Phillips as a serial interconnect for complex integrated circuits. It has been widely used in mass-produced consumer electronics applications and a large variety of low-cost integrated circuits (from Phillips Components and other manufacturers) are available for connection to the bus. Use of this bus enables the pin count of integrated circuits to be reduced, which, in turn, simplifies the interconnect wiring patterns required to be implemented on PCBs used to mount such integrated circuits. The maximum bit rate that can be used to transmit messages on this bus is about 100 kb/s. The I^2C bus is a multimaster bus with collision detection and arbitration procedures to prevent data corruption if two or more masters simultaneously initiate data transfer.

A block diagram of a typical I^2C bus system is shown in Figure 8.6. Two bus lines are required: a serial data (SDA) line and a serial clock (SCL) line. Figure 8.7 shows the wired-OR method used to connect devices to the bus. Each integrated circuit connected to the bus has a unique address, and a master/slave protocol is used. In a multimaster system masters can operate as master–transmitters or as master–receivers. A master device generates clock pulses for the SCL line. Slave devices receive commands and clock signals from a master via the bus. Each integrated circuit designed for the bus has its own unique address determined during manufacture. Some integrated circuits allow part of the address to be set by external control. A seven bit address is used with an eighth bit (LSB) used to control read and write (if this bit is

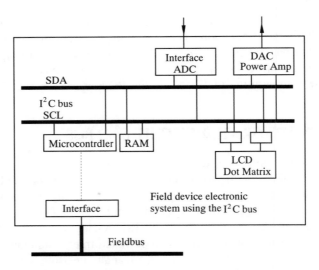

Figure 8.6
Block diagram of I^2C bus system

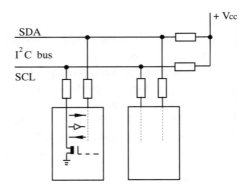

Figure 8.7
I^2C wired-OR connection method

1 a master device reads data from a slave device, and when it is 0 a master device writes data to a slave device). Integrated circuits available for use on the I^2C bus includes microcontrollers, RAM, EE-PROM, I/O interfaces, infrared transmitters and receivers, ADCs and DACs, and real-time clocks.

The bus protocol is activated by the SCL and SDA signals. Table 8.1 shows the basic transitions required to control the bus.

Valid data is defined when SDA is high and SCL passes through a L–H–L transition. The first byte transmitted after the start condition contains the 7 bit

8.3 INTER-INTEGRATED CIRCUIT (I^2C) BUS

Table 8.1
Basic bus control operations for the I^2C bus

SDA	SCL	Bus State
H	H	Quiescent
H → L	H	Master start
L → H	H	Master free

integrated circuit address and the R/W bit. The addressed integrated circuit responds by returning an ACK pulse. If the R/W bit of the address byte is 0 the master sends data to a slave until it no longer receives ACK pulses, or until all the data has been transmitted. If the R/W bit is 1 the master generates clock pulses which enables the slave to send data. After every received byte the master generates an ACK pulse. This brief description of a transaction on the I^2C bus provides an indication of the simplicity of the protocol.

Interfacing the I^2C bus to the parallel parts of integrated circuits is an important operation that enable devices not specifically designed for the bus to be connected to an I^2C system. The PCD8584 is an example of an I^2C controller that performs this interface function. Figure 8.8 shows the block diagram of this device.

Other serial interfaces for board-mounted devices have been developed. The four wire bus systems introduced by Motorola, National Semiconductor, Hitachi, Texas Instrument and Maxim are briefly discussed below.

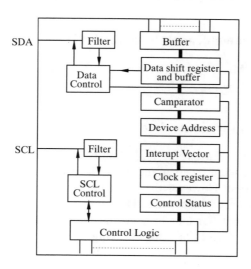

Figure 8.8
Block diagram of the PCD8584; an example of an I^2C controller integrated circuit

Motorola serial peripheral interface (SPI)

The four lines of this bus are defined as follows:

- SCK: serial clock
- SS: slave select, from decoded address lines (this input does not determine if a device is a master or a slave)
- MISO: master in, slave out (if the slave has no digital output, e.g. as for a DAC the MISO line can be deleted)
- MOSI: master out, slave in.

In the basic protocol, the master (typically a microcontroller) selects the slave by: dropping SS, clocking an 8 bit frame by driving SCK and MOSI while simultaneously receiving 8 bits via MISO, then raising SS to end the frame. Clock phase and polarity are software-programmable in the master. Most SPI slaves determine clock phase on the fly by latching the clock level when SS becomes low.

National semiconductor microwire

This is similar to SPI and it is also based on an 8 bit frame. In Microwire SCK latches data on the rising edge only, while SPI allows either clock pulse or polarity. Also, a variable number of bits per word are possible in Microwire.

Hitachi serial communication interface (SCI)

This is similar to Microwire. SCI shifts data LSB first, rather than MSR first, as in SPI and Microwire.

Texas instrument serial interface

This is used in the TMS 320 DSP family. It is more flexible than SPI. Completely independent receiver and transmit sections allow for either half- or full-duplex operation. However the clock latching edges for data in and data out are the opposite of Microwire. Data out may or may not be present in devices which do not transmit back to the microcontroller; if available, it is used for cascading devices or error checking.

8.3 INTER-INTEGRATED CIRCUIT (I^2C) BUS

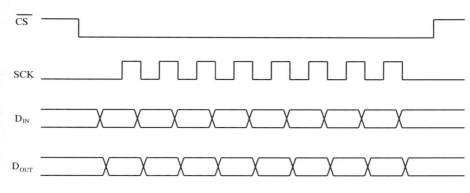

NOTE: D_{IN} CLOCKED IN ON SCK RISING EDGE.
D_{OUT} CHANGES ON SCK FALLING EDGE.

Figure 8.9
Waveform protocol of the Maxim serial interface bus

Maxim

This serial interface bus is fully compatible with Microwire and it is compatible with SPI, provided that the SPI register control bits are set to CPOL = 0 and CPHA = 0. It is compatible with the Texas Instrument bus, provided that an inverter is included on the SCK lines and hardware compatible with the Hitachi bus but bit order requires significant software data manipulation to achieve compatibility. It is compatible with all microcontroller port lines toggled under software control. A diagram illustrating the waveform protocol of the Maxim serial interface bus is shown in Figure 8.9.

9
The Connected Future

9.1 SERIAL NETWORKS

This book is concerned with a small part of the serial networks scene. Serial communication techniques in fact impact technology at all levels; from the small scale of integrated circuits, through large-scale factory and process automation, to worldwide communication networks. In the future the web of serial communication links will be totally connected with the field network at the lowest level of the hierarchy of information technology networks.

A field serial network can be multimedia involving; twisted pairs, coaxial cables, optical links, fibre optic cables, power lines and wireless links. They must have defined data link and application layer functions and in many cases users will require a defined user layer. Serial field networks are a major enabling technology for the implementation of all types of automation schemes. More specifically they are a major enabling technology for measurement and control engineering. The low cost and high capability of software and electronic systems combined with serial network architectures enable a distributed data base approach to be adopted for the design of measurement and control instrumentation. This reduces the dependence on a central system (and hence improves reliability) and, reduces wiring and installation costs.

All of the conceivable automation areas will be affected in some way by the development of serial field networks. Each market area (e.g. industrial (process and factory), domestic buildings and equipment, commercial buildings and office equipment, laboratory, medical and transport) has one and, in many cases, several serial field networks. Some networks are now being sold for general applications and this no doubt introduces a range of marketing and customer support problems.

Very little appears to be have been done for the entertainment or education application areas, although the Musical Instrument Digital Interface (MIDI) system (Rothstein, 1992) has been used successfully for many years. Allik, Dunne and Mulder (1990) propose a standard network for real-time control and communication between performance systems such as music synthesizers, tape players, lighting equipment, slide, laser and video projectors, and live performers. It was envisaged that this network, called Arconet (Artists Computer Network), would overcome the limitations of the MIDI system. However it does not appear to have progressed beyond the proposal stage.

Emerging applications such as multimedia interfacing will motivate interest in satisfying the requirements of the entertainment and education application areas. Multimedia interfacing requires low cable costs and real-time performance. A new serial bus, the IEEE P1394 High Performance Serial Bus, has been specially designed for this purpose. This bus will offer 100, 200 and 400 Mb/s data rates, and it will use a six-wire cable comprising two shielded twisted pairs (for data and control information) and two power supply wires. It will feature:

- node branching and daisy-chaining topologies
- 'live' cable insertion and removal
- automatic reconfiguration whenever a node is added or removed from the network
- memory-mapped architecture
- a serial bus protocol which will define how SCSI functions will be transported over the bus.

Application areas envisaged for this new bus include the interconnection of computers, video and audio equipment, digital cameras, hard disk drives, CD-ROMs, printers, scanners and consumer electronics.

In this book the main automation application areas for serial field networks have been considered in Chapters 4–7; factory automation and process control, laboratory and medical instrumentation, intelligent buildings and transport, respectively. The specific standard networks are discussed separately in Appendix A, and Appendix C discusses three of the available integrated circuit protocol implementations. This approach has enabled the design context of each network to be established, along with a discussion of their main technical and performance features. A limited discussion of the electrical features of serial network technology has been presented in Chapter 2. The reader should consult one of the many texts discussing digital communication networks if detailed information about the networks and protocols is required. Chapter 8 discussed the use of serial networks in electronic instrumentation with emphasis on the way that physically small field units can be constructed when these techniques are adopted. Chapter 3 presented a general discussion of network standards and emphasized the importance of the standardization process for product development. In many cases standardization becomes an intimate part of the product development sequence. This final chapter addresses future developments and provides concluding comments.

9.2 STANDARDS AND PRODUCTS

It has been common practice to base product differentiation and market share on the competitive advantages of proprietary technology. The objective of this strategy was of course to lock customers into a particular product range. To combat this strategy, users will choose products with a wide supplier base. However in the case of network-related products users may be attracted to a dominant supplier, since co-ordination by a single supplier may be more effective than by means of vendor or standards writing groups. As a result of the difficulties experienced when information technology products are interconnected, the trend at the present time is towards open systems defined by a standard.

An open system uses hardware and software defined by standards. This should allow software applications and information to be easily moved to systems manufactured by different organizations and will facilitate the implementation of the portability feature desired by the majority of users. Interoperability is another widely desired feature that will allow equipment of different size and origin to co-exist on a serial network. This will be achievable at some level but the inclusion of options in the defining standards will require strict control. Since options usually result from the vested interests of standards committee members some level of Government intervention will be required to constrain the number of options and increase the level of achievable interoperability. It should be noted that if interchangeability is offered as a feature of a particular field network instrument, then it can be replaced by a similar instrument from another vendor and the system will instantly return to normal operation. Interoperability is normally seen as a higher-level requirement, since instruments offering this feature can be replaced as described, but they will also offer additional functionality without affecting the normal operation of the system.

Integrated circuit technology enables the efficient implementation of complex standard communication protocols. It also enables convertor circuits to be developed which can be used to link different networks. This will lead to equipment that will allow networks to co-exist and may be the evolutionary path that will be taken towards a single network defined by a standard acceptable to a wide range of market sectors. It will inevitably be a pre-product standard and it will be important to ensure that the innovative smaller company is able to contribute to the standardization process.

The higher layers of a serial network communications link are very important for the success of the network. Users will want to configure their systems by means of a software system that will allow them to use a diagram or natural language interface. Vendors designing instruments for the process control industry have been particularly active in this area. A standard notation is required to describe field devices to be connected to the bus. The HART device description language (DDL) has been adapted for fieldbus applications. It has recently been submitted (under the name 'virtual device syntax') for inclusion in a layer 8 definition for the IEC Fieldbus standard. Function blocks have also been defined for fieldbus applications. They can have input and output data flows that can communicate with other function blocks and they have mode and alarm features. Typical functions include control function blocks (e.g. PID); they will enable a variety of control strategies, including single input/single output feedback loops, cascade, feedforward and ratio control. Functions can be distributed among sensors and actuators. If designed correctly, interfaces such as the device description language and function block languages will make the serial network technology transparent to the user.

A definitive test for conformance to a standard will be difficult to establish since, as discussed previously, non-conformance can only be accurately identified. This is an important issue, since legal problems associated with the product reliability and safety defects of a product will be resolved by reference to the conformance of a product to a standard (or a set of standards). Conformance testing should demonstrate that a product implements a standard correctly and establish its ability to interoperate with other conformant products. A commonly accepted interoperability test methodology has not as yet been established. It is unlikely that conformance testing will provide information defining the robustness and reliability of a product. Serial network standards are complex documents, often with many options, so procedures for testing a product for conformance must be highly automated to reduce the cost of testing and the time devoted to testing. Conformance testing is an engineering activity that will become increasingly important.

This book has stressed the increasing impact of standards on the development of serial networked field instrumentation and product innovation in general. Standards writing has become a pre-product activity with the standard becoming part of the product definition phase. It is important to recognize that an involvement with a standard is not necessarily altruistic: for example, a large investment in the standards process could be made simply to influence the outcome or to acquire the leading edge knowledge of the committee. Because of cost and time constraints, standards committees are often dominated by the larger commercial organizations so some form of Government support for the smaller company or academic organization to participate is required. All staff levels, from management teams to research engineers working on the initial definition stage of a product, should be involved. It will be necessary for academic organizations to respond to this changing situation by including an overview of appropriate standards and, standards writing and use at a much higher level in engineering courses. Measurement and control standards

9.3 SYSTEMS INTEGRATION

cover a wide range including EMC, quality control, software and communication methods. In this book attention has been restricted to standards defining the communication protocols and connecting media used to interconnect field instrumentation.

9.3 SYSTEMS INTEGRATION

The human interface is of major importance for the success of any system. User-friendly is the usual phrase used to describe what is perceived to be a system that is easy to use, without requiring excessive memorizing of instruction sequences or constant reference to system handbooks. Interface design has received considerable attention over the last decade (Booth, 1990) and can be expected to be increasingly important. Complex serial networked field instrumentation introduces three major human interface levels. At the highest level a user can be imagined as operating a system with little knowledge of the system other than that presented by the interface workstation. At some lower level, technically but maybe not physically, closer to the field devices a workstation will be used to initially set-up the system and then be subsequently used to provide supervisory and maintenance information. This intermediate level will provide the user with unrestricted access to the detailed operation and design of the system. Of course these higher-level interfaces are simply software configurations, so access will in fact be controlled by a password. At the lowest, field device, level a handheld terminal will be used at installation and under fault conditions. An essential feature of a serial networked system with intelligent field devices is that it will appear to be a distributed data base. Consequently the storage requirements of the central system will be significantly reduced with device information actually stored in the field device. This should lead to the elimination of many volumes of user information and greatly improve the usability of the system.

Weiser (1991), in an interesting paper discussing computer developments for the 21st century, comments that 'The most profound technologies are those that disappear. They weave themselves into the fabric of everyday life until they are indistinguishable from it'. Such a disappearance is a fundamental consequence, not of technology but of human psychology. Whenever people learn something sufficiently well, they cease to be aware of it. Weiser envisages hundreds of computers (devices with embedded computers) in a factory or office using small and large range wireless connections, and very high speed wired connections. He was concerned with information processing, so his devices included handheld/clip on units, lap-top (notepad) units and display boards (hightech black boards), all incorporating computing units that will operate in a transparent manner. This amounts to a comprehensive decentralization of computing power. Clearly this approach can be applied directly to serial networked field instrumentation and will lead to a similar decentralization of measurement and control instrumentation.

Who will be responsible for the user interface? Each market will have its own user layer. This is well recognized by the process control and factory automation vendor, but even here the application layer of the IEC Fieldbus standard is making slow progress towards completion. The application layer is the important link between the application (the user) and the protocol and the physical bus. In intelligent commercial buildings where a range of proprietary serial bus systems are used there is probably less scope (or demand) for a user to add unexpected devices to the bus so the user interface should be relatively easy to control. Intelligent domestic buildings appear to present more problems because of the potentially large diversity of products and vendors. A separate organization, acting as the system integrator with responsibility for the user interface, could well be required here. In any event, in all of the market areas it is likely that the role of the system integrator will be increasingly important. The system integrator will be responsible for ensuring that the technology is transparent to the user, that the interface is user-friendly. He will also be responsible for the initial installation of this system and he will be the point of contact for the user when fault conditions arise.

The system requirements at the initial installation stage are similar to the requirements of instrumentation vendors interested in rapidly prototyping serial networked field devices. The major requirement here is an ability to connect a new device to a bus when it is live, followed by a rapid and automatic reconfiguration of the bus to take account of the new device. The use of a fieldbus core device as a platform for rapidly prototyping sensor or actuator systems will make a valuable contribution to product development groups working in the measurement and control area.

9.4 FLEXIBLE MANUFACTURING

Modern manufacturing organizations must be able to respond quickly to market changes. A company survives by introducing automation to enable it to be flexible, offering a variety of products with a range of options. A specialized plant for each product line may require less sophisticated control and handling equipment, but it is then difficult to change. A flexible manufacturing system provides a range of shareable resources which can be selected and deployed under computer control. Each operating station in this type of system is likely to include sub-systems for computer numerical control (CNC), workpiece transport control (WTC) and tool support control (TSC). Figure 9.1 shows the typical network hierarchy that is commonly used to control a production line based on

9.4 FLEXIBLE MANUFACTURING

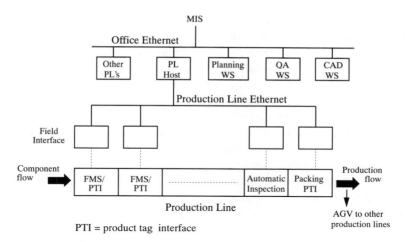

Figure 9.1
Flexible instrumentation for flexible manufacturing systems

flexible manufacturing systems. Clearly the FMS is most conveniently implemented as a manufacturing version of a distributed data base system. Currently the field interface is based on programmed logic controller (PLC) technology.

A PLC is basically a specialized microcomputer system that collects data from plant equipment and issues control signals. PLC functions were originally carried out by hard-wired relay structures. The modern PLC replaced these by software control of electronic logic systems. PLCs are usually operated with a supervisory control and data acquisition (SCADA) system. PLCs will be used to control fieldbus systems. It is interesting to speculate that a fully populated fieldbus can be viewed as a distributed PLC, so it is conceivable that once a full range of intelligent field devices becomes available the need for the PLC will diminish considerably.

At the lowest level of the FMS network hierarchy fieldbus systems will in many cases be required to interface with the product as it moves along the production line or is moved between production lines by an AGV (automatic guided vehicle). Manufacturing automation is making increasing use of bar codes and radio frequency automatic identification tags to replace the use of paper documents to keep track of a product as it moves through the factory. Cohen (1994) describes these techniques and the large number of 'standard' codes and communication protocols that have already been developed. Some degree of unification with the IEC Fieldbus standard is clearly desirable but at the time of writing appears to be unlikely.

Data tags are attached to each product or to the support arrangement that is used to transport the product. An RF transponder is mounted at each assembly station. When a product arrives at an assembly station the manufacturing sequence required at that point in the production cycle is read from the tag. These manufacturing operations are then executed and the results of these operations are then written to the data tag. Hence a manufacturing history is built-up in the data tag and travels with the product and in some cases will remain with it when it is shipped to the customer. Hence a centrally stored software map of product progress as it passes along the product line is not required. This has the effect of reducing the flow of large blocks of data over the production line serial network and allows the use of slower networks with master–slave protocols.

AGVs are themselves complex systems which will benefit from the application of fieldbus techniques to their own instrumentation systems. Like product tags, AGVs will be required to communicate with the fieldbus network by means of radio or optical wireless links. AGVs navigate by means of wire following or optical path following systems. Gyro-based and optical image processing systems have also been proposed for AGV navigation. AGV communication protocols are covered by a separate standardization group.

The rapid prototyping feature of fieldbus systems discussed in the previous section may find useful application in flexible manufacturing systems. For example it could provide an ability to quickly configure (or re-configure) an automatic measuring system or to add sensors for condition monitoring or quality assessment. The use of intelligent field devices may also further reduce the need for skilled labour to supervise the modification of a flexible manufacturing system. The design of networked field devices for rapid prototyping requires further consideration.

A flexible manufacturing system effectively eliminates the common mode failure problems associated with a centrally controlled system. Failure of mechanical production line units remains a problem area that is contained by regular maintenance. Increasingly condition monitoring techniques (e.g. for example involving vibration, proximity and vision sensors) are being used to support an as required maintenance policy. Condition monitoring sensors are essentially wide-bandwidth devices, and extensive signal processing must be used before information from the sensor can be reduced to a much lower data rate and transmitted over a fieldbus. Fortunately low-cost, low-power and small-size integrated electronic systems are available for sensor interfacing. This is another area that requires further consideration.

9.5 CONCLUSIONS

This book has described the wide range of serial field device networks that have been developed for each market sector where measurement and control instrumentation can be found. Serial networks have been presented as an

9.5 CONCLUSIONS

enabling technology for the automation of large- and small-scale manufacturing, process control and monitoring systems. It is this technology that will allow the low-cost implementation of automation schemes. The inevitable impact of this technology will be to reduce further the level of human involvement in the manufacturing process. Fortunately it is also this technology that will facilitate low-cost small product run automation of manufacturing processes. It will enable small run variety and quality to be obtained using mass-production techniques. Hopefully this will lead to an increasing range of employment opportunities.

Access to communication over a serial bus will open-up considerable opportunities for sensor (and actuator) manufacturers to add value to their products. The additional measurement of ancillary variables, and the measurement of the significant parameters of high-frequency noise for fault-detection purposes can be noted as two possibilities. In addition calibration control via the bus will be possible with minimum intervention by maintenance personnel. A close involvement with integrated circuit technology will be required if low-cost serial network devices are to be produced.

A uniform information technology approach to system automation, from management level down to field devices, will provide benefits to vendors and users. Vendors restricted to producing communication devices for a particular market sector are clearly little influenced by the work of vendors working in other market sectors. A single integrated protocol circuit covering the majority of application areas is desirable and feasible. High-level software satisfying user requirements in each market sector will always be a distinctive requirement. The trend towards instrumentation vendors supporting more than one serial bus system is increasing and this will probably lead to an increasing need for integrated protocol conversion systems. It remains to be seen if the IEC fieldbus will be able to overcome the competition of established proprietary networks.

Appendix A
Standards

A.1 INTRODUCTION

Thirteen serial networks are described in this Appendix. A further significant number of networks can be recognized if industrial proprietary bus systems and the variations in the home automation and vehicle application areas are included. Each distinct market area uses one (and often several) serial bus systems. These market areas can be broadly identified as follows:

- military systems
- civil aviation
- vehicles (road automation)
- intelligent (commercial) buildings
- domestic automation
- process control
- factory automation
- laboratory instrumentation
- medical instrumentation

Many common features can be recognized so it is not surprising that bus systems are being developed to cover a wide range of application areas (for example, the proprietary Echelon LONWORKS and the IEC fieldbus). Table A.1 lists the major serial bus systems and notes the coding method used, access methods, ISO conformance, availability of integrated circuits and the type of standard (e.g. defined by a trade association (TA), national standard (N), proprietary standard (P), and international standard (I)).

In this Appendix brief notes are presented on each of the major serial bus systems. Additional information will be found in the chapters discussing the application areas for serial networks linking field devices.

A.2 ARCNET

This first appeared in 1977 as a local area network for use in office applications. Integrated circuits have now been developed that allow a considerable reduction

Table A.1
The major serial networks designed for field instrumentation applications

Network	Digital Code	Access	ISO Layers	Integrated Circuits	Standard
Arcnet	NRZ	Token passing	1,2	SMC	TA
ARINC 629	Manchester	CSMA	1,2	MCE	TA(SAE)
CAN	NRZ	CSMA	1,2	Intel, Phillips Motorola	P
CEbus	NRZ	CSMA	1,2,3,7 (user language)	?	N(EIA)
Echelon LON	Manchester (differential)	CSMA	1,2,3,4,5,6,7 (user language)	Neuron chip (Motorola/Toshiba)	P
FIP	Manchester	Master/slave	1,2,7	VLSI Technology	N
IEEE 1118 (Bitbus)	NRZI	HDLC	1,2,7	Intel (Bitbus)	N
IEEE P1073	NRZI	Master/slave	1,2,7 (user language)	None available	N
IEC Fieldbus	Manchester	Token passing plus master/slave	1,2,7*	Ship Star	I
MIL-STD-1553B	Manchester	Master/slave	1,2	Wide range of devices available for military use	NATO
PROFIBUS	Manchester	Token passing plus master/slave	1,2,7	ISP	N

*standard not complete
SMC Standard Microsystems Corporation
MCE Micro Circuit Engineering
ISP Interoperable Systems Project (Ship Star, 1994)

in circuit complexity and have opened a wider range of applications in industrial automation. The ARCNET standard specifies baseband signalling at 2.5 Mbits/s and a token passing protocol for media access to a maximum of 255 nodes with a maximum data packet of 512 bytes. Two network topologies are specified namely star topology and bus topology. The centre of the star functions as a wiring concentrator (commonly called a hub) which repeats the incoming signal from any radial element of the star onto all other radial elements. The wiring concentrator may amplify or regenerate the physical signals but it does not interpret or modify the signals. The star topology is effectively a collection of point-to-point connections between the hub and remotely located stations. These point-to-point connections may be used to link hubs and to mix bus and star topology segments to form a single network.

The standard defines the physical layer, a data link layer (comprising media access control and logical link control sub-layers) and network management

A.2 ARCNET

functions. A media access control sub-layer provides sequential access to the shared bus medium by passing control of the medium from station to station in a logically circular fashion. This sub-layer determines when the station has the right to access the shared medium by recognizing and accepting the token from the predecessor station and determines when the token shall be passed to the successor station. Upon receipt of the token the station may perform an enquiry phase and a token transfer phase, prior to relinquishing control of the medium by performing the token transfer phase. If a station receives a token when it has no information to transfer over the medium, that station immediately enters the token transfer phase.

The token loop is automatically configured upon network initialization and reconfigured dynamically as stations enter and leave the network. The network reconfiguration process is invoked whenever the token is lost. The previously active station leaves the network or stays and becomes active on the network. This reconfiguration process involves each station determining the address of its successor station on the token loop by polling of sequentially ascending station addresses. Once the address of the successor station is determined, tokens are passed directly to that station without further polling until the next network reconfiguration. In cases where the token has been lost or a new station has become active on the network, any existing network activity is forced to halt and the reconfiguration process is initiated. A time-out procedure based on station address is used to select the active station with the highest address for the purpose of starting the token loop.

The timers defined in Table A.2 are used at each station to control various operational characteristics of the network. Several of these timer values are fixed, while several others are variable and must be set to equal values at all stations on the network. The variable timer values are referred to in terms of extended time-outs. Support for extended time-outs are optional, but if supported all extended time-outs must be selectable.

The term reset when applied to timers is to be understood to mean that the time is reset to its initial value and restarted. When a timer's interval is expired it is said to have timed-out, which asserts a time-out condition which remains true until the timer is reset. A timer may be stopped prior to time-out in order to prevent its time-out conditions from occurring.

When network reconfiguration is needed, as detected by power on reset conditions or the TLT time-out at the station, the station activates its transmitter to transmit a reconfiguration burst, the reconfiguration burst does not communicate useful information and is used to force network initialization by terminating all activity on the network. Network activity may be terminated in this manner because the reconfiguration burst is longer than any type of frame and will therefore interfere with the next invitation to transmit frame. By interfering with the invitation to transmit the frame, the reconfiguration

Table A.2
ARCNET timers

Timer		Specification	Comments
Lost Token	TLT	840 ms ± 10 ms	Reset each time the station receives the Token and it is used to initiate network reconfiguration when a time-out occurs before the next token is received.
Identification Precedence	TIP	TOL0, K = 146 µs TOL1, K = 584 µs TOL2, K = 1168 µs TOL3, K = 2336 µs TIP = K(255 − ID) + 3 µs	Provides time separation for initiation of network reconfiguration activity based on the station address.
Activity Time-out	TAC	TOL0, 82.4 − 87.6 µs TOL1, 329.6 − 350.4 µs TOL2, 659.2 − 700.8 µs TOL3, 1318.4 − 1401.6 µs	Used to control the minimum period of time which the station will wait for bus activity before assuming that such activity will not occur and hence commence network reconfiguration.
Response Time-out	TRP	TOL0, 75.6 − 77.6 µs TOL1, 302.4 − 310.4 µs TOL2, 604.8 − 620.8 µs TOL3, 1209.6 − 1241.6 µs	Used to control the minimum period of time which the station will wait for a response to a transmitted ITT, FBE or PAC frame before assuming that such response will not occur.
Recovery Time	TRC	2.0 µs ± 0.4 µs	Provides time separation between the end of a response time-out and the start of a token pass.
Line Turnaround	TTA	12.0 µs to 13.6 µs	Controls the minimum interval between the end of a received transmission and the start of a transmitted response.
Medium Quiescent	TMQ	4.0 µs to 4.8 µs	Controls the sampling interval used to determine if a transmission is taking place on the bus.
Receiver Blanking	TRB	5.6 µs to 6.4 µs	Controls the interval after the end of a transmission that the receiver is to be blanked before valid network activity can be received.
Broadcast Delay	TBR	15.6 µs to 20.0 µs	Controls the minimum interval between the end of a broadcast transmission and the start of a token pass.

burst prevents any station from receiving the token. This will ultimately result in a TAC and/or TLT time-out at all stations.

The active hub implements cable segment, isolation, wiring concentration, signal repeating and echo cancellation. Active hubs which provide features such as network management or media type conversion may be implemented as a super set of the functionality defined here. The active hub imposes a priority upon its ports based upon the first port from which it detects incoming activity during any particular signal repeating event. Assigning a fixed set of relative priorities among these ports is acceptable to resolve incoming activity conflicts which occur due to detecting input signals simultaneously on two or more ports. However, a strict first-come first-served priority must be used to arbitrate between activity arriving in sequence on different ports.

An N-port active hub employs a finite state machine with $N + 1$ states. One of these states is an idle state when the receivers are enabled on all ports and the transmitters are disabled on all ports. As soon as activity is detected on any port a transition is made on the other N states based upon the port on which activity was first detected. The resulting state is a state in which the receiver of the port from which the activity was detected remains enabled, while the receivers of all the other ports are disabled; and with transmitters enabled on all ports other than the port from which the activity was detected. Once this active state is entered, all signals received on the (enabled) receiver port are retransmitted approximately simultaneously on all other ports.

Note that in the ordinary circumstance this regeneration of the signals does not involve retiming of the signals; therefore bit jitter accumulates from hub to hub, thereby limiting the number of hubs which may be used on any path between stations.

A.3 MILITARY STANDARDS 1553B AND 1773 (MIL-STD-1553B AND MIL-STD-1773)

Work to define a serial bus standard for electrical cables used in military avionic applications started in 1968 and this led to MIL-STD-1553A in 1975 and its updated version MIL-STD-1553B in 1978. Equipment based on these standards has proved to be flexible and very reliable. MIL-STD-1553B has been adopted for use in all types of military equipment, including avionics, ships, tanks and some missiles.

A fibre optic version of MIL-STD-1553B was published in 1987 and later became MIL-STD-1773. This was a natural progression since the reduced weight and natural EMI immunity of fibre is an attractive feature in many applications. A shielded 1553B bus can offer very high levels of noise immunity but this is achieved at the expense of increased weight. The 1773 standard

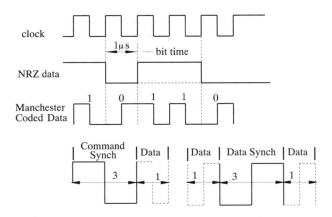

Figure A.1a
Manchester code and synchronization patterns specified by 1553B standard

was designed to enable a fibre optic system to replace a 1553 electrical system with no impact on existing equipment other than requiring the replacement of electrical transmitters and receivers with optical devices.

The MIL-STD-1553B bus is a time division multiplexed, half-duplex communication link using a 1 Mb/s data rate. Figure A.1a shows the Manchester code and synchronization patterns specified by the standard. A twisted pair cable (with coupling transformers) links a maximum of 32 nodes or instrumentation sites. One of the nodes on the bus must act as a bus controller. Information is transferred in a command/response mode, with the bus controller (BC) having complete control over communication and with no node speaking until spoken to. A 1553B node is called a remote terminal (RT) in the standard document and has a unique address on the bus. Each RT has self-test facilities, and fault tolerance can be enhanced with the use of an optional redundant bus.

The MIL-STD-1553B was published before the ISO communication model was developed. It covers the physical layer and data link layer of the ISO model. The application layer functions are completely undefined and must be

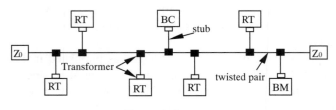

Figure A.1b
1553B bus topology

A.3 MILITARY STANDARDS 1553B AND 1773

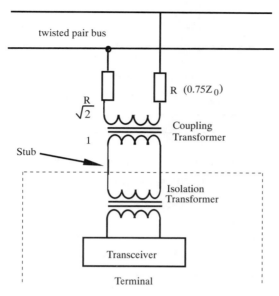

Figure A.1c
Transforming coupling specified by 1553B standard

supplied by users. A companion handbook MIL-HNBK-1553 contains guidelines for implementing application layer functions such as message scheduling and error control.

Figure A.1b shows the basic 1553 bus structure. Redundant bus structures are also defined by the standard. Transformers are used to link stubs and devices to the twisted pair bus as shown in Figure A.1c. Three basic devices are defined; namely, bus controller (BC), bus monitor (BM) and remote terminals (RT). The bus controller is effectively a master device and all other devices are slave devices. The bus monitor records bus activity which provides diagnostic information and allows the bus monitor to act as a back-up controller since it will contain the information about the current status of the other terminals essential for transfer of bus control functions. Remote terminals interface remote instrumentation to the bus. It should be noted that simple terminals can implement a subset of the available message formats and a terminal can flag a message as illegal if it is not configured to accept that message. This allows terminals with a wide range of complexity to be connected to the bus, and simple terminals need not carry the overhead of implementing unnecessarily complex messages.

There are three different word types defined in the 1553B standard, namely, command, status and data words. Figure A.1d shows the structure of these

Command word

bit times					
3	5	1	5	5	1
Command synch	Terminal address	T/R	Subaddress/Mode	Word count/Mode count	P

Status word

bit times											
3	5	1	1	1	3	1	1	1	1	1	1
Command Synch	Terminal address	ME	I	SR	Reserved	BC	B	S	DB	T	P

ME = Message Error
I = Instrumentation
SR = Service Request
BC = Broadcast Command received

B = Busy
S = Subsystem
DB = Dynamic Bus control acceptance
T = Terminal

Data word

bit times		
3	16	1
Data synch	16 Data Bits	P

Figure A.1d
Command, status and data words defined by 1553B standard

words and the meaning of typical bit patterns. Information is transferred over the bus in the form of messages constructed by combining these words as shown in Figure A.1e. All messages start with a command issued by the bus controller. If an RT is to respond then it should send its status word followed by any data words required.

The number of data words to be transmitted or received is specified in the original command. The data must be continuous with no gaps between words and the RT must respond to the command within 4–12 bit times. The 1553 standard defines a bit time as 1 μs. There must always be a gap of at least 4 bit times between consecutive messages.

A high noise immunity results from the use of shielded twisted pair cable, Manchester coding, transformer coupling and relatively large voltage levels. Additional security is provided by checks made by the data link layer to assess the validity of a message. All terminals on the bus must be capable of performing these checks and any message which does not pass them is ignored, with the result that the bus controller will not receive a status word and will assume that an error has occurred. Checks are performed at bit, word and message level. Error checking is based on a single parity bit in each word, which must comprise 16 data bits and an odd parity bit, and each bit is checked

A.3 MILITARY STANDARDS 1553B AND 1773

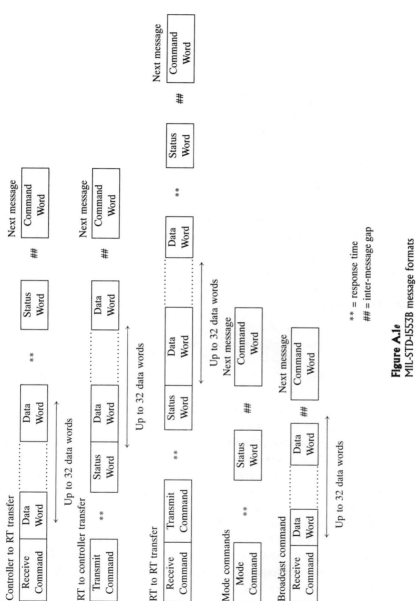

Figure A.1e
MIL-STD-1553B message formats

for a valid Manchester waveform. If any single bit or synchronization pattern is invalid then the entire message is deemed invalid. Each message is checked for the correct number of words (specified in the initial command word) and that it is continuous with no gaps between words.

A.4 IEEE 1118

This standard describes a serial bus interconnecting microcontrollers or any programmable electronic system. The intended broad applicability of this standard is evident in the application areas listed in the standard document:

- instrumentation
- test and measurement
- distributed data acquisition systems
- industrial automation, (a) manufacturing control and (b) process control
- building control systems, (a) residential, (b) commercial, (c) environmental, (d) access and intrusion
- distributed I/O, (a) RS 232 type peripherals, (b) data entry and display devices, (c) sensors and actuators, (d) control devices and (e) human interface devices.

A multidrop bit-serial communication protocol is defined with messages not greater than 256 bytes (a typical message having 128 bytes) and signalling rates selectable in the range 50–500 kb/s. The IEEE 1118 communication model has a sparse compliance with the OSI model. Network, transport, session and presentation layers are null and industry independent bus services (generic bus services) are provided at the application layer. System management functions are specified for all of the defined layers.

Two types of physical media are defined by the standard. Physical layer twisted pair type A is used when compatibility with existing equipment is needed, operation at 62 kb/s (with maximum cable length of 300 m) and 375 kb/s (with maximum cable length of 1200 m) and a hardware reset is needed. Physical layer twisted pair type B is used when the electrical environment is noisy and media redundancy is needed. Type B media operates at 76.8 kb/s and 268.8 kb/s, which are multiples of the standard communication rates, to facilitate the design of instruments that operate using both the 1118 standard and conventional area network standards. The binary information on the physical media is encoded by the non-return-to-zero inverted (NRZI) method. All information, except for the beginning and closing flags, obeys the HDLC/SDLC rule of zero bit insertion (i.e. when there are five consecutive ones, input a zero bit in the data stream on transmission and remove the

A.4 IEEE 1118

inserted zero bit on reception). The maximum number of consecutive ones is six, and it occurs in the beginning and closing flags. This zero bit insertion and deletion is automatically performed within most HDLC/SDLC communication interpreted circuits.

The data link layer of this standard is based on the HDLC protocol. Significant differences include a pre-frame synchronization field added in front of the opening flag to facilitate clock recovery, and the achievement of a Hamming distance of three is facilitated by a period of no transitions between frames providing strong frame delimiting.

The data link layer provides three classes of data-transfer services and uniform interface for upper-layer entities to use in accessing these services. Datagram and acknowledged datagram connectionless services permit a message to be sent, or messages to be exchanged, without requiring a connection to be established. A connection-oriented service enables more reliable data transfers which is obtained by establishing synchronization between two communicating devices and provides for missing and duplicate message detection and, in cases of corrupted or missing messages, data retransmission. A third service, the device control service, provides a mechanism to identify and test devices connected to the bus. In addition the data link layer provides broadcast (multicast to all stations) and generalized multicast (group addressing) services.

The data link layer performs an error control function and controls redundant media (an operational feature of this standard bus). The error detection provided is sufficient to achieve an undetected error rate that is less than or equal to one error every twenty years and a Hamming distance of three.

Application layer entities that support general-purpose operation are defined by the generic bus services. These services are designed to support typical microcontroller implementations which require services such as reset, download, upload and message passing. A summary of the services provided by the generic bus service is as follows:

- device control services, (a) reset devices, (b) mark device offline, (c) read device information, (d) enable/disable memory space access, and (e) get/set time
- memory access services, (a) upload/download data, (b) upload/download code, (c) read/write scratch pad memory and (d) read/write special function registers
- I/O access services, (a) read/write I/O location and (b) logical I/O (AND, OR, XOR, SET)
- task access services, (a) create/delete task, (b) suspend/resume task, (c) get task function ID, (d) read task function IDs, and (e) task message exchange

- support for user extensions, (a) define services, and (b) call service.

The system management layer of the 1118 standard defines functions required for interaction with layers 1, 2 and 7 of the communication model. These functions include:

- activation/deactivation functions, (a) device reset, activation, and deactivation, (b) device identification and logical address assignment, and (c) selection of media
- bus mastership control, (a) selection of a bus master at power-up, (b) take over of mastership by alternate master after failure of active master, (c) passing of mastership under user control, and (d) ability to delegate bus mastership for a specified length of time under user control
- monitoring functions, (a) device status, (b) status of the bus master and alternate masters and (c) network performance statistics
- error detection and recovery, (a) error statistics, (b) network diagnostic and (c) switching of media after failure.

At its simplest the 1118 standard bus can be configured as a single-level hierarchial network with a master controlling several slave units on a multidrop bus. A multiple level hierarchy is constructed by arranging for the slaves to have an associated sub-master which is used to control another multidrop bus.

A.5 ARINC 629

Two protocols are defined by this standard: the basic protocol (BP) and the combined mode protocol (CP). The basic protocol has two operating modes: periodic and aperiodic. Buses designed to operate in the periodic mode automatically revert to the aperiodic mode when the bus is overloaded. The essential feature of the combined mode protocol is that it allows periodic and aperiodic messages to be transmitted on the same bus. Bus access is based on the CSMA/CA method and transmitted signals are Manchester encoded with a bit rate of 2 MHz. Current mode and voltage mode transmissions are specified over unshielded twisted pairs. A fibre optic bus option is specified.

Periodic data are data that appears on the bus at regular fixed intervals. The periodic mode is usually the preferred mode when the bus is carrying signals whose frequency content must be maintained (for example for servo-control signals). The ARINC 629 standard defines a transmit interval (TI) so that if the time needed for all terminals to transmit once, including the bus overhead, is less than the TI, then the bus is said to be operating in the periodic mode. By using a TI value small enough such that the sampling rate is greater than the highest frequency data, the frequency content of the data can be preserved.

A.5 ARINC 629

Aperiodic data is generated asynchronously and updated at a non-uniform rate (for example as in position reporting for landing gear systems). Often some degree of data latency is tolerable. However it should be noted that the use of aperiodic data minimizes the bandwidth requirements for a bus at the risk of increasing the data latency.

Broadcast data transfer and directed data transfer methods are specified by the ARINC 629 standard. The broadcast message is sent without acknowledgement. It is the most efficient data transfer method. The directed transfer (or point-to-point transfer) method uses specific labels and control commands to designate the sender and receiver to control the transfer using acknowledge and clear-to-send commands.

Equipment is connected to the physical medium through a data bus terminal. Each terminal is programmable, and up to 120 terminals may be connected to a single bus. Terminals monitor the bus and wait for a quiet period before transmitting. In each terminal, three programmable timers are used to ensure that the terminal transmits at the proper time on the bus. Messages on the bus are composed of wordstrings, with a maximum of 31 wordstrings in a message. Wordstrings begin with a label word, followed by zero to 256 data words. Receiving terminals decode the label and determine if the data following it are required of the receiving system. Checksum and CRC error detection methods are used. The polynomial to be used in the CRC algorithm is $P(x) = x^{16} + x^{12} + x^5 + 1$.

The basic protocol uses three rules to ensure orderly transmissions, without collisions or contentions (for periodic or aperiodic operation). A terminal cannot transmit again until a TI has elapsed, a quiet period called the sync gap (SG) has existed on the bus, and a quiet period known as the terminal gap (TG) has elapsed on the bus since the sync gap occurred. TI is the longest time of the three timer periods. The SG timer is the second longest and it is reset if a carrier appears on the bus before it has elapsed. Once it has elapsed it should only be reset when its terminal begins transmitting. The TG timer is reset by the presence of a carrier. The TG time interval should begin only after the SG period has elapsed and only if no carrier is present. Unlike the SG, once it has elapsed the TG is reset by the presence of a carrier. The TG and SG should not overlap in time, they should run consecutively. The periodic and aperiodic operating modes are not exclusive. A minimum frame time (MFT) is defined such that when it is less than TI the bus is operating in the periodic mode. Alternatively, when the MFT is designed to be larger than the TI, the bus is operating in the aperiodic mode, and in this case transmissions may still be periodic, but this cannot be guaranteed.

The combined mode protocol allows for substantial levels of aperiodic data to be transmitted without affecting periodic data transmission. If aperiodic demand transiently exceeds the available bus time, the protocol automatically limits and queues the aperiodic traffic to fully utilize the bus capacity and maintain message priorities. The combined mode provides three levels of bus access priority. Level 1 allows periodic transmissions of nominally constant length messages. To ensure access in any TI for all terminals at level 2, it is necessary for the bus to be designed with sufficient available bus time, after the bus level 1 load is accounted for. The TI can be selected from the range 0.5 to 64 ms. Level 3 message transmission is in ascending order of the TGs of those terminals wishing to transmit. Level 3 is the bus access level intended for the bulk of the aperiodic messages.

A clip-on inductive coupler is specified for unshielded twisted pair bus media. Typically a linear bus is formed with stubs connecting the bus coupler to devices and equipment. The main features of the cable specification are: characteristic impedance 130 Ω, capacitance 30 pF/m with 30 twists/m. Manchester encoding of signals is specified. Increased efficiency can be obtained by further encoding the Manchester signal by arranging for a short duration AC signal (a doublet) to be transmitted at each change of state of the Manchester signal. This enhanced encoding method reduces the low-frequency content of the transmitted signal, while the high-frequency signal content is unchanged.

A.6 PROFIBUS

This fieldbus was sponsored by the German Federal Ministry for Research and Technology (BMFT). It is defined by the German national standard DIN 19245 part 1 and part 2, and it conforms to three layers of the OSI communication model; namely, physical layer, data link layer and application layer. It was designed to offer a standard open communications interface for simple and complex field devices and provide a simple interface to MAP networks. Layer 7 of MAP is defined by ISO IS 9506 (the MMS protocol) and layer 2 and 3 of PROFIBUS is organized so that fieldbus functions can be transferred to MMS functions. In addition to MAP the PROFIBUS specification is based on the EIA-RS 485 (ISO DP 8482) physical medium standard and the IEC 955 standard (Process Data Highway, Type C — PROWAY C).

A half-duplex asynchronous transmission method is specified with UART characters defined by DIN 66022/66203. The transmission medium is specified to be twisted pair cable with transmission rate from 9.6 to 500 kb/s. A maximum length of 1200 m is specified without repeaters and 4800 m with repeaters. The number of stations is limited by the address range to be 127 maximum with 32 of these allowed to be complex, master, devices (see below). Physical address assignment is optional and an address is valid as a global address or broadcast address.

A.6 PROFIBUS

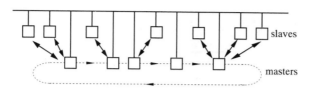

Figure A.2a
PROFIBUS token ring

The data link layer of PROFIBUS defines a bus access method that combines the master/slave method with the token passing method. Field devices can be active (limited to a maximum of 32) and passive. Active stations can be masters and a token ring can be established to link active stations. A station holding the token can communicate with any of the other stations using a command/response method. Figure A.2a shows a diagram of a token ring linking masters which, when they hold the token, can communicate with slaves using the command/response method. Note that to satisfy the sampled data and other timing requirements of measurement and control applications it is necessary that the token be passed from master to master within a defined time frame.

PROFIBUS offers four data transmission services for cyclical send and request operations, namely: send data with no acknowledgement (SDN), send data with acknowledge (SDA), request data with reply (RDR) and send and request data (SRD).

SDN is used for broadcasts from an active station to all other stations on the bus and therefore an acknowledgement is not appropriate. SDA and RDR are the essential elementary services of a command/response protocol and SRD reduces the overhead of bus activity when a combination of SDA and RDR can be achieved. Note that slaves are able to communicate with masters by placing data into the immediate response message of an incoming SRD or RDR service.

To satisfy cyclic access requirements PROFIBUS enables a polling list to be created in the data link layer. Cyclical polling is based on the acyclic RDR and SRD services. These cyclic services are called CRDR (cyclic request data with reply) and CSDR (cyclic send and request data). The services of the data link layer interface with the application layer are as shown in Figure A.2b.

This diagram also shows the fieldbus network management which enables, for example, the station address, data rate and station status to be transferred to the data link layer.

The PROFIBUS application layer consists of two layers; namely, the lower layer interface (LLI) and the fieldbus message specification (FMS). The LLI is primarily responsible for flow control (the function of ISO Layer 4, the Transport Layer 1), establishing and terminating connections, and conversion

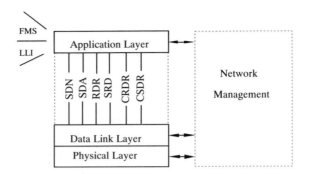

FMS = Fieldbus message specification
LL I = Lower layer interface.

Figure A.2b
Services of the data link/application layer interface

of data link layer functions to the fieldbus message specification (which is essentially the interface to MMS). Communication between different stations is performed through communication relationships (CRs). LLI distinguishes CRs into one-to-one and one-to-many/all CRs and handles them. One-to-one CRs are connection-oriented, one-to-many/all CRs are connectionless, i.e. multicast or broadcast. Connection-oriented CRs can exist between two master stations (master/master connection) or between one master and one slave station (master/slave connection). They are distinguished into connections for cyclic or acyclic data transfer. Some services can be mapped on both connections for cyclic or connections for acyclic data transfer. For each connection an attribute has to be set in the configuration phase of the system to define this connection as cyclic or acyclic. Cyclic data transfer is performed through the layer 2 services CSRD or CRDR. Acyclic data transfer is carried out through SRD (master/slave connections) or SDA (master/master connections).

Multicast or broadcast data transfer requires the establishment of no connection (connectionless CRs). Only unconfirmed data transfer (SDN) is possible through connectionless CRs. Therefore the source station gets no information as to whether or not the sent message was correctly delivered to the destination station. Multicast or broadcast data transfer can be used, e.g. for clock synchronization or global alarms.

Each application process contains a certain amount of objects. Objects (e.g. variables, program code segments, data segments) are parts of the application process. The application process can allow application processes in other stations to gain access to a set of its own objects (e.g. to read variables). This particular set of objects is termed VFD (virtual fieldbus device). The following objects are defined within the VFD: domains, program invocations, datatypes, variables, variable lists and events.

One application process (the client) can access objects of application processes (the servers) in other stations by using FMS services such as: initiate and conclude, reject, status, get object description, create object, delete object, read, write, information report, event notification and event acknowledgement. Some FMS services are allocated a priority parameter.

These services are derived from manufacturing message specification (MMS, ISO 9506) to support the interconnectivity to higher automation levels. They are optimized for application needs in the field. To interpret the incoming data a database with descriptions of the objects is necessary. This database is termed the object dictionary. One entry in this object dictionary describes an object completely. Certain objects (e.g. variable lists, program invocations, etc.) can be created dynamically. Their description must be added to the object dictionary. Addressing an object is done using an index. This mechanism guarantees fast access to objects.

A.7 FIP: THE FACTORY INSTRUMENTATION PROTOCOL

FIP is an open field bus network defined by a set of French National Standards (UTE — 46 — 6 xx, see Figure A.3*a*). It conforms to three layers of the OSI communication model; namely, the physical layer, data link layer and application layer. FIP is not just a serial network to interconnect field-located instrumentation; it is also a system for updating and managing a distributed data base.

The physical layer defines a maximum bus length of 2000 m with a maximum of 256 stations. The bus cable is specified to be twisted pair (with a fibre optic option) and the data rate of the Manchester coded signal is specified to be 1 Mbit/s with options of 31.25 kbit/s and 2.5 Mbit/s. Figure A.3*b* shows

General architecture	C 46-601
Physical layer	
Copper	C 46-604
Optical fibre	C 46-607
Data link layer	C 46-603
Application layer	C 46-602
Application layer	
MMS subset	C 46-606
Network management	C 46-605

Figure A.3*a*
The set of French national standards that define the factory instrumentation protocol (FIP)

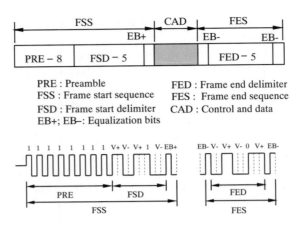

Figure A.3b
Message formats and coding specified by the FIP standard

how the control and data information are framed by preamble and delimiter bit patterns.

The data link layer of FIP offers two kinds of basic services: transmission of variables with critical time services and message services with MMS compatibility. Media access procedures are based on the PDU model (discussed in Section 3.3). A bus arbitrator (BA) performs the function of the distributor. The BA offers three operating modes; namely periodic traffic and aperiodic traffic for variables and for messages. For periodic traffic the BA calls variables by their names according to an order defined by a table. For aperiodic traffic the BA calls variables upon request from any station or the BA transmits an 'emission right' to the station requesting access to the media (each variable is broadcast over the network by the transmitter concerned and is accepted by all users requiring the variable). Messages can be transmitted to another station (with or without acknowledgement) or to several stations (broadcasting without acknowledgement). Any FIP station can perform the bus arbitration or the variable production/variable consumption functions, but at any given time only one station can be an active bus arbitrator.

In the periodic mode the BA scans a table containing a list of identifiers to be circulated on the bus. A question frame (ID_IDAT) is broadcast and is recorded by the data link layers of all stations connected to the bus. One station responds as the producer of the identifier and one (or more) of the stations recognize that they will be consumers of the variable when it appears on the bus. The producer broadcasts the variable corresponding to the identifier in a response frame (RP_DAT) and this is accepted by all of the consuming stations. This procedure is repeated as the BA scans the table. When the system is configured the BA is given the list of variables to be scanned and their associated periodicities.

Only one BA is active at a given time. The others monitor the bus activity and if the active BA fails, a local mechanism is implemented in all potential BAs to select a new active BA.

The application layer provides for real-time selection of scanning cycles, confirmation of quality control and variable transmission, coherence of distributed data base elements, and synchronized sampling (by sensors) and synchronized commands (by actuators). It also offers a sub-set of the MMS services without interfering with real-time traffic. The application layer offers a set of services to the application process which allows it to read or write the values of variable which may be integer, floating point, boolean, etc., or be an array or a list, etc., and may have a range of different descriptive names, e.g. variable, synchronization variable, consistency variable.

A local reading and writing service interfaces a station to the network, but although local writing and reading requests are initiated by producers and users, respectively, local reading and writing operations are independent of network activity. Remote reading and writing are controlled by the PDU access control method. The time status of the read and write operations is also managed by the applications layer. For example, a production delay counter sets a status flag indicating that a refreshment status is true indicating to a consumer that the information produced is valid. Another delay flag is set by promptness status and indicates to a consumer that the consumed variable has been refreshed by the network within a prescribed time. Consistency operations are implemented which indicate whether the network has repeated the distribution periods of each item of a list of related variables.

A.8 HART

HART is an acronym for highway addressable remote transducers (Bowden, 1991). It is an open proprietary standard originated by Rosemount. The HART specification defines the physical form of transmission, the transaction procedure, message structure, data formats, and a set of commands. The physical form of HART is based on the industry standard 4–20 mA analogue signalling method with the bottom 4 mA used to provide power and a superimposed current modulation providing the means for digital communication.

Figure A.4a shows the basic multidrop connection method and Figure A.4b shows the typical hardware required to couple to a microprocessor-based field device.

Frequency shift keying (FKS) is used to code digital information. A 0.5 mA peak sine wave is used with logic 1 represented by 1200 Hz and logic 0 represented by two cycles of 2200 Hz; this gives a data rate of 1.2 kb/s. The

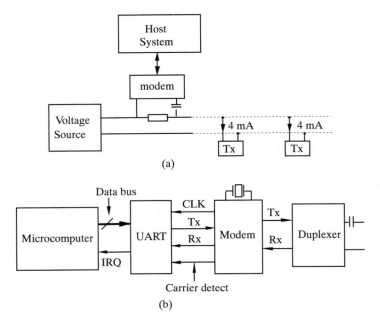

Figure A.4
Block diagram showing the use of the HART protocol implementing the multidrop bus connection method (*a*) and a block diagram showing its basic hardware requirements (*b*)

average current of this modulating signal is zero so it can be used (if required) with a fully functioning 4–20 mA signalling method. A widely used telephony standard modem, the Bell 202, enables the easy implementation of this digital signalling method.

The HART protocol is a master–slave protocol where a field device only responds when it receives an instruction via the bus. There can be two masters (for example, a controlling (host) system and a hand-held interface). The field device's status is included in every reply message, indicating its continued good health. Two or three message transactions can be made every second. The basic message structure is shown below:

PREAMBLE	see below
STRT	start character
ADDR	address — source and destination
COMM	command (Table A.3 shows the typical commands)
BCNT	byte count
STATUS	device and communication status (2 bytes)

A.8 HART

Table A.3
Typical HART protocol commands

Universal commands
    Read manufacturer and device type.
    Read primary variable (PV) and units.
    Read current output and percent of range.
    Read up-to four pre-defined dynamic variables.
    Read or write 8-character tag, 16-character descripter, data.
    Read or write 32-character message.
    Read transmitter range, units and damping time constant.
    Write polling address.

Common-practice commands
    Write damping time constant.
    Write transmitter range.
    Calibrate (set zero, set span).
    Set fixed output current.
    Perform self-test.
    Perform master reset.
    Trim PV zero.
    Trim DAC zero and gain.
    Write transfer function (square root/linear).
    Write sensor serial number.

Device-specific commands
    Read or write low flow cut-off value.
    Start, stop or clear totalizer.
    Choose primary variable (mass flow or density).
    Read or write material or construction information.
    Trim sensor calibration.

DATA   up to 24 bytes
CHK   check sum

Each byte of information has the following parameters: one start bit, eight data bits, one odd parity bit and one stop bit.

Every message is preceded by a specified number of hexadecimal FF characters called preambles. The preambles enable the receiver's signal detection circuit. As part of the response to the initial query sent by the master, each field device will inform the master of the minimum number of preambles that it requires for satisfactory operation. The master then sends that number of preamble characters in all subsequent transactions with that field device. Since the master cannot know this number until the field device responds, the initial query should send 20 preamble characters (alternatively average throughput is increased if the minimum (most probable) number of preambles is sent on the first attempt). If there is no response, the retry will default to the maximum of 20 preambles. The protocol uses a 3-byte pattern recognition method to indicate the start of a message. The message detection patterns are FF FF 02 for the master transmitted message and FF FF 06 for the field device transmitted message.

Once a message detect pattern has been recognized the detection circuit is disabled until the end of the message.

The checksum byte contains the exclusive-OR of all the bytes which precede it in the message, starting with the start character. This provides a further check on transmission integrity, beyond that provided by the parity check on the 8 bits of each individual byte. The combined effect of these parity checks allows the detection of any single burst of up to three corrupted bits in a message.

A.9 CEBUS AND THE HOME BUS SYSTEM

The CEbus network (Fitzpatrick and Markwalter, 1989; and Evans, 1991a) is a tree structure consisting of nodes and routers. The nodes are the terminal devices that communicate with each other for interproduct status monitoring and control. Routers connect different wired media and provide the packet forwarding function between the wired media. Devices providing a similar function to routers are used to connect infrared (or radio frequency) media to the wired media. The network structure is dynamically reconfigurable; the system regularly updates itself to maintain the tree structure. The routers transmit topology packets around the network which allows routers to detect if there is a loop in the network. The router that is the cause of a loop will deduce that it is the problem and stop forwarding all packets. This breaks the loop and maintains the tree structure. The benefit of the tree structure is that only one path exists between any source/destination pair of nodes, which allows two simple routing methods: directory routing, where routers forward packets only if this progresses them towards their destination; flood routing, where every router forwards the packet along all branches of the network.

The CEbus is defined by four layers of the OSI communications model. The three basic layers (physical, data link and application) are used, and in addition the network layer is specified to facilitate multimedia operation. The transport, session or presentation layers are not explicitly used. The physical layer specification includes six different physical media, namely; power line, coaxial cable, twisted-pair (Evans, 1991a; Khawand *et al.*, 1992), infrared (Hofman, 1991), radio frequency (Winick, 1991), optical fibre. A transceiver for the twisted pair bus has been described by Dickson (1991). The twisted pair medium consists of four twisted pair cables; this is compatible with the joint

A.9 CEBUS AND THE HOME BUS SYSTEM

EIA/Telecommunication Industry Association telephone premise wiring standard (EIA/TIA — 570). This allows equipment based on the CEbus standard to be retrofitted into existing homes within the American market. The CEbus uses non-return-to-zero, pulse width encoding with four symbols: 1, 0, end of frame and end of packet.

The CEbus protocol is based on a packet message format containing control fields and information fields (see Figure A.5a). The control fields dictate how a message is to be delivered across multiple media to reach its destination. The information field contains the standards application language, the Common Application Language, CAL (Khawand et al., 1991), that provides the application interface for the CEbus.

The bus access method (Markwalter et al., 1991) is based on CSMA with contention detection and contention resolution (CSMA/CDCR). The detection of an earlier signal on the line prevents a node from interrupting the transmission of other nodes. The contention avoidance schemes employed by the CEbus are:

- prioritization (high, standard and deferred)
- queuing
- randomization of start time delay interval within each priority.

Data units define the information exchange between peer layers of the communication model. For the CEbus the application layer protocol unit (APDU) comprises a header followed by a command from the Common Application Language (CAL) that has been specified for this bus. This basic structure is used for the other layers.

The CEbus application layer is the user interface to the network. It supports the CAL through which users communicate with devices connected to the bus. The application layer is responsible for dividing long messages into shorter segments that can fit into one CEbus frame.

The CEbus network layer provides connectionless, acknowledged and unacknowledged services. It adds a header to the APDU to form the network

> Preamble
> Control field
> Destination node address
> Destination home address
> Source node address
> Source home address
> Information field (up to 32 bytes)
> Checksum

Figure A.5a
CEbus packet format

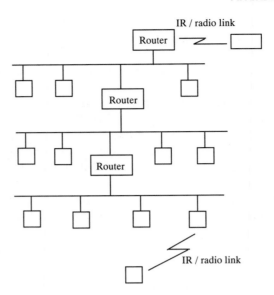

Figure A.5b
Routers used in the CEbus multimedia network

layer protocol data unit (NPDU). Routers, which link the different physical media of the CEbus (as shown in Figure A.5b), do not require an application layer, so the network layer is the highest layer in this case. As mentioned above, two types of routing are supported: directory routers and flood routers. It is interesting to note that the IEC International Fieldbus standard, which will eventually specify the use of twisted pair, wireless and fibre optic networks, does not include a network layer.

The CEbus data link layer is responsible for implementing frame assembly and disassembly, error detection and medium access control. It provides point-to-point acknowledged and unacknowledged or broadcast unacknowledged connectionless transfer. A header (called a control field in this case) is added to the NPDU to form a logical link protocol data unit (LPDU). This function is performed by a logical link control (LLC) sublayer. A media access control (MAC) sublayer adds preamble and checksum fields to form the CEbus frame that is translated by the physical layer into the electrical signals that propagate over the bus.

Japanese work in the area of serial networks for home automation applications has closely paralleled the work of the CEbus groups in the USA. Muratia *et al.*, (1983) presented an early proposal for the standardization of Home Bus Systems. Subsequently, in 1988 the Electronic Industries Association of Japan established the Home Bus System Standard (ET-2101). This standard specifies

Priority code
Source address
Destination address
Control code
Byte count
Data area (Header code, source subaddress, destination subaddress, source subdevice address, destination subdevice address, operation code/operation data) Frame check code

Figure A.5c
The Japanese HBS control channel packet format

only the layers 1, 2 and 3 of the OSI communication model. The *IEEE Transactions on Consumer Electronics* (1983–1993) has published the results of the work of the Japanese group.

The HBS standard specifies a multimedia network. A 9.6 kb/s bit rate is specified and access control uses the CSMA/CD method with priority and bit contention. The HBS protocol supports four priority levels. If two colliding packets have the same priority level then arbitration passes to the source address bits. This will resolve the collision, as the colliding packets will not have the same source address. The control channel format is shown in Figure A.5c. An information channel is specified for video and, voice and data.

A.10 ECHELON LONWORKS

This is a proprietary local operating network (LON). The Echelon product includes a special-purpose integrated circuit, called the Neuron chip, and extensive software to support design and operation of networks. It implements all seven layers of the OSI reference model, as shown in Table A.4. The access method uses a modified version of the carrier sense multiple access (CSMA) media access control algorithm called predictive p-persistent CSMA which has been designed to enhance the media access sublayer, allowing improved performance for multiple-media communication over large networks with heavy bus traffic.

To simplify routing, the LonWorks protocol defines a hierarchical form of addressing using domain, subnet and node addresses which can be used to address an entire domain, subnet or an individual node. The protocol also defines another class of address called groups, which enable the addressing of multiple dispersed nodes.

Four basic types of message are supported. These are acknowledged, request/response, unacknowledged repeated and unacknowledged. Acknowledged messages are the most reliable but for large numbers of nodes they impose a greater load upon the network than the other types. Messages can

Table A.4
Echelon LonWorks protocol layers

OSI layer	Services Provided	Purpose
Application	Standard network variable types	Application compatibility
Presentation	Network variables Foreign frame transmission	Data Interpretation
Session	Request–response authentication Network management	Remote actions
Transport	Acknowledged/Unacknowledged Unicast/Multicast Authentication Duplicate detection	End-to-end reliability
Network	Addressing routers	Destination addressing
Link	Framing/Data encoding Predictive CSMA Error checking	Media access and framing
Physical	Multimedia interfaces (twisted pair, power line, radio frequency, coaxial cable, infrared, fibre optic)	Interconnect

also be given a priority level to improve the transmission time. This is achieved by assigning a priority slot to a node. If a priority packet is generated within this node, it is sent ahead of any pending non-priority packets buffered for transmission. When it reaches a router or bridge, it goes to the head of the message queue, behind any other queued priority packets.

For security purposes, this protocol supports an authenticated message service. This allows receivers of a message to determine if the sender is authorized to send the message. This is achieved using a common 48 bit authentication key. When the message is sent, the receiver challenges the sender to provide authentication using a random challenge which is different each time. The sender then uses the authentication key to perform a transformation on the challenge before responding. The receiver compares this response to the challenge with its own transformation on the challenge. If they match then the transaction succeeds. The transformation used is designed so that it is extremely difficult to deduce the authentication key, even if the challenge and response are both known. The use of this service will slow down the transmission rate of this bus system.

Nodes are connected in local operating networks (LONs) using a protocol called LonTalk. The LON moves short sense and control messages unlike a local area network (LAN) which is designed for large data transfer situations. A LON can transmit a maximum of 1.25 Mb/s which translates into more than 500 messages/s.

A.10 ECHELON LONWORKS

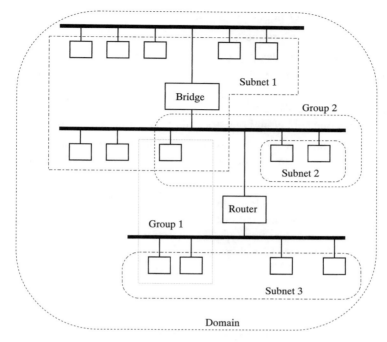

Figure A.6
A LonWorks topology

The LonWorks system consists of domains which can be divided up into subnets and groups, as shown in Figure A.6. A domain is a logical collection of nodes on one or more channels and communication can only take place between nodes in the same domain. A domain effectively forms a vertical network. Multiple domains can occupy the same channels, therefore they may be used to prevent interference between nodes in different networks that share the same channel. Within a domain there may be subnets, which are smaller logical connections of nodes. All nodes in a subnet must be on the same channel or on channels connected with bridges. Subnets cannot exist across routers. If a node is configured to belong to two domains then it must be assigned to a subnet within each one of the domains.

A group is also a logical collection of nodes within a domain but the nodes are grouped together without regard for their physical location in the domain. A node can be configured to be a member of up to 15 groups. Groups are an efficient way in which one node can address a large number of nodes in a short time interval. Within each domain LonTalk has the following restrictions:

- a maximum of 225 subnets

- a maximum of 127 nodes per subnet
- a maximum of 256 groups
- a maximum of 64 nodes per group (this restriction applies for acknowledge services only. There is no node limit on groups using unacknowledged services)
- a minimum of 32 385 nodes
- a node has one subnet and one node address per domain to which it belongs
- a node may have up to 255 network variables defined (echelon has defined 50 variables (e.g. time, voltage, current and speed) with their units, range and increment).

The Neuron chip is the basic building block of the Echelon system. It is part of each node in a LON and allows equipment at these nodes to communicate with each other. There are two types of Neuron chip available to the consumer. These are the Neuron 3120 chip and the Neuron 3150 chip. The Neuron is discussed in Appendix C.3.

A.11 IEEE P1073

The medical information bus is defined by a family of three standards: IEEE 1073.1, 1073.2 and 1073.3. It is based on the ISO Open System Interconnect.

1073.1 deals with the general architecture and logical components that must exist throughout an MIB implementation. It covers the upper three OSI layers: application, presentation and session layers. The session layer provides structural and synchronized control of the communication between a device and the host. Attributes of a session connection, termed a token, are dynamically assigned. These tokens allow control to be exercised for data transmission, connection and disconnection. ISO 8326/27 specifies basic connection-orientated service definitions and protocols for the session layer. The presentation layer is concerned purely with the representation of data values for transfer. Because it is possible to represent data using any number of bit patterns, an abstract syntax is defined to delineate a set of primitive elements whose range of values is fully outlined. This allows a formal definition of the information to be transferred with minimum constraints on the representation of the data. The application layer comprises a set of common application service elements (ASEs) which effectively open and close connections, and transfer data. A well-designed application layer will have a strong influence on user satisfaction with equipment based on the 1073 standard.

The types of data operations that can be invoked are defined by the medical device data language (MDDL). The MDDL communicates measurement data, status and control information between the host system and the medical instruments. Clinical data involves much more than simple numerical values. Extra parameters such as measurement units, medical device status and information concerning the method and time of measurement also have to be taken into account.

The MDDL describes each message as containing two sentences: transactive sentences and active sentences. The transitive sentence contains information identifying the source and destination device and also message type. It contains three types of messages: command, query and report, each of which is concerned with changing, inquiring and describing current settings or states. An example is shown in Table A.5a.

In the first line the host sends a query for data to a pump device. In the second line a monitor sends the host a report of the data that was requested in line one.

The active sentences contain the data. The grammar is designed to allow applications to convey both the content of the data and the context in which it was obtained. It is very detailed in its descriptions, which allow it to handle the flow of information around the medical bus. An example is shown in Table A.5a.

Table A.5b shows an example of three MDDL active sentences. In the first, an infusion pump device specifies an infusion rate of 50 ml/h for a report. In the second, the systolic blood pressure (aliased BPS to save on network traffic) is reported to be 120 mm Hg. The final line is a shortened version of line 2.

MDDL is an object-orientated language. It treats devices, host computers and parameters as objects and provides methods for describing and processing messages between these objects. The parameters of new devices will automatically be added to the MDDL whenever these tables are updated allowing a flexible method for standardized host–device communication. IEEE P 1073.2

Table A.5a
MIB transactive example

{Source}	{Verb}	{Destination}	{Type}
Host	Sends	Pump	Query
Monitor	Sends	Host	Report

Table A.5b
MIB active example

{Alias}	{Subject}	{Discriminant}	{Fundamental}	{Verb}	{Units}	{Datum}
	Device		Infusion rate	Equals	ml/h	50
BPS	Patient	Systolic	Blood Pressure	Equals	mm Hg	120
BPS				Equals	mm Hg	120

defines the bedside communications subnet, and corresponds to the lower four OSI layers: the physical, data link, network and transport layers. The primary goal is to provide fast, reliable communication while also automatically detecting device connections (or disconnections).

The physical layer of 1073.2 uses RS485 with non-return-to-zero inverted (NRZI) encoding at 375 kbps. This was chosen for its high noise immunity. A unique 6 pin connector based on the AMP SDL connector is specified, comprising 3 pairs of wires and a shield. One pair is used for half-duplex communication. The second carries a timing pulse to synchronize instruments. The last carries ±12 V DC with current limited to 250 mA. Thus power is provided to the device communications controller (DCC) interface for devices with limited power.

The data link of 1073.2 provides the service primitives for connection-orientated point-to-point data transfer. The bedside communications controller (BCC) functions as the master station and all DCCs function as slave stations. The BCC employs a polling protocol to detect device connections (or disconnections). It polls each port with a frequency of 1 to 10 Hz. Connection requests are negotiated through the exchange of exchange identification (XID) packets. Maximum information field length is negotiated through the use of MIB loading units. The BCC polls empty ports by sending XID frames that offer the maximum amount of loading units available to that port. The connection request at the DCC specifies the loading units needed by the device. If the units offered are adequate, the DCC sends a XID response frame in response to the XID poll. Strict response times and transmission retry limits are specified. Failure to respond to a BCC poll or the absence of a poll at the DCC within a 1 s period breaks the connection and returns the port to disconnected status.

The network layer of 1073.2 is the MIB inactive-network (MIN). The reason why it is inactive is that no routing is needed in a point-to-point configuration, however it does have service primitives for connection, data transfer and disconnection. These primitives are present to allow for possible future expansion while maintaining backwards compatibility.

The 1073.2 provides transport layer services between a DCC and BCC. The MIB inactive transport (MIT) layer service primitives are similar but not compatible with ISO 8073 transport services. MIT does not support packet assembly/disassembly, but it does contain a simplified set of active parameters in its connection service primitives.

A.12 CONTROLLER AREA NETWORK, CAN

CAN uses the NRZ (non-return-to-zero) format for transmitting its data. Although this method exhibits good noise emission characteristics and it is

A.12 CONTROLLER AREA NETWORK, CAN

very efficient for encoding bits, the network synchronization could be lost if a large number of identical bits are transmitted. To solve this, CAN employs bit stuffing with the stuff width specified at five bits. Bit stuffing stops the loss of synchronization by inserting a bit of the opposite polarization to the current data stream every n bits, where n is the stuff width. The receiving nodes should then be expecting this, allowing them to resynchronize to this bit and then discard it. The stuff width is very important for the success of the network's operation: a small stuff width will lock the synchronization firmly in place, but the number of stuff bits will increase dramatically, leaving less time for the actual message. The opposite is also true: a large stuff width will gain message time, but lose the network's stability. Note that a small stuff width would also reduce the bit encoding efficiency that makes NRZ a better choice than RZ (Return to Zero) encoding.

It should be noted that bit stuffing can lead to errors. If a bit is corrupted on the bus it could cause the appearance of a stuff bit in the data field of the message. At the receiver, the bit would be destuffed and the data could be corrupted. However Bosch have analysed this bus and found that the probability of undetected errors during the lifetime of a vehicle is extremely low.

CAN has two types of data frame: a standard frame and an extended frame. A standard frame message consists of eight data fields in the following order:

SOF, ARB, RTR, CTRL, DATA, CRC, ACK, EOF

SOF is the start of frame message, which consists of a single dominant bit and is used to synchronize the receiving nodes together.

ARB is a 12 bit arbitration field which consists of an 11 bit identifier (ID) and the RTR bit (see below). An 11 bit identifier is capable of addressing 2 or 2048 possible message destinations. The value of the ID selects the priority of the message, with low values of ID having high priority.

RTR is technically part of the ARB field. It is a single bit which indicates whether the message should have a reply or not.

CTRL is a 6 bit control field. It uses 4 bits of information to determine the number of bytes; there will be in the data block, an IDE bit used to indicate whether the frame is a standard or an extended one (it is transmitted dominant for a standard frame) and a reserved bit (transmitted dominant) which can be used if the system is expanded in the future.

DATA is a data field which consists of between 0 and 8 bytes of data. The exact number of bytes will have already been determined in the CTRL field.

CRC is a 16 bit error checking field. The first 15 bits are used to check for errors in the data, and the 16th bit, called the CRC delimiter, is transmitted recessive. The CRC check covers the SOF, ARB, RTR, CTRL, DATA and CRC fields.

ACK is a 2 bit acknowledge field. The first bit is called ACK-slot and the second is called ACK-delimiter. These bits are transmitted recessive, and if any node wished to acknowledge reception of the data, then it should transmit a dominant bit to overwrite ACK-slot, which will then be detected by the original transmitter. Note that ACK-slot is surrounded by two recessive bits (CRC-delimiter and ACK-delimiter), making the acknowledgement easy for the CAN controller to recognize.

EOF is a 7 bit field used to signal the end of the data frame. The bit stuffing process is deactivated during this field, as there is no need to keep strictly synchronized to it.

The extended data frame is exactly the same as the standard frame, except for the following.

The ARB field has 32 bits instead of 12. These bits are split into an 11 bit base ID which is equivalent to the standard field's ARB ID, and an extended ID of 18 bits. The two IDs are separated by an SRR (substitute remote request) bit and the CTRL field's IDE bit.

The SRR bit is a recessive bit transmitted at the position of the RTR bit in standard frames.

The IDE bit is transmitted recessive, indicating that this is an extended frame.

The CTRL field's IDE bit is replaced with a reserved bit. Both reserved bits are transmitted dominant.

The extended frame can be used with the standard frame. They are compatible with each other on the same network. CAN ICs capable of using the extended frame format can also transmit using the standard format, but ICs which were designed to use the standard format only cannot receive messages in extended format. Should an extended frame and a standard frame with the same 11 bit ID be arbitrated against one another, the standard frame will gain the bus, due to the extended frame's recessive SRR bit.

The two main implementations (Full CAN and Basic CAN) can use either a high-speed or a low-speed network. The two speeds require different network configurations and bus voltage levels. The performance characteristics for low speed CAN are summarized below:

- up to 125 kbit/s bit rate
- maximum bus line length is determined by capacitive load (the number of nodes on the network) and the bit rate; there may be between 2 and 20 bus nodes
- the network is a two-wire differential bus line with a common return line

A.12 CONTROLLER AREA NETWORK, CAN

- the network is short-circuit proof from -6 V to $+16$ V ($+32$ V for 24 V powered vehicles); the common mode range is -2V to $+7$V; a single $+5$ V power supply is required
- the transmitter output current > 1 mA
- there is a single termination network for each bus wire.

The performance characteristics for high speed CAN are summarized below:

- 125 kbit/s to 1 Mbit/s rate
- the network bus may be up to 40 m in length, running at 1 Mbit/s bit rate
- there may be between 2 and 30 nodes on the bus
- the network consists of a two-wire differential bus line with a common return line; the characteristic line impedance is 120 Ω
- the network is short-circuit proof from -34 V to $+16$ V ($+32$ V for 24 V powered vehicles); the common mode range is -2 V to $+7$ V; a single $+5$ V power supply is required
- transmitter output current > 25 mA.

Traditional master/slave and token-passing protocols have not been adopted for in-vehicle applications since it is claimed that they cannot be designed to provide the necessary timely response (with predictable and minimum latency) to the dynamic and asynchronous actions of the vehicle. CAN is a development of the CSMA/CD access method which uses non-destructive bit-wise arbitration to allow messages transmitted simultaneously to be arbitrated in an prioritized manner. A transmitting node monitors the bus, hence it is able to detect when it loses arbitration and is then to revert to a receive mode until the higher priority message finishes its transmission. A feature of CAN that also contributes to its timely performance is its message acknowledgement schemes. As a result of these schemes all nodes are aware of the status of each message, data consistency is assured between nodes, and the recovery time from errors may be kept below 25 μs when the bus is operating at 1 Mb/s.

CAN was developed by Bosch for in-vehicle applications. It has been found to be more generally useful and an international users and manufacturers group, CIA (CAN in automation), has been formed. This group promotes the use of CAN in industrial applications and develops proposals to make CAN product comply with the full ISO seven-layer communication model.

Bosch currently licenses several semiconductor companies to develop integrated circuits for CAN applications (e.g. Intel, Phillips and Motorola). In addition a CAN bus driver is available from Phillips to replace the large number of discrete components previously necessary to implement the physical layer media interface.

A.13 IEC FIELDBUS (IEC DIS 1158)

The standard defining this fieldbus will be an eight-part series of documents. It will be significantly larger than any of the other serial network standard documents. The eight parts are, from 1 to 8; Introductory Guide, Physical Layer Specification, Data Link Service Definition, Data Link Protocol Specification, Application Service Definition, Application Protocol Definition, Fieldbus Management and Conformance Testing. The Physical Layer Specification is approved for publication, the Data Link Layer (parts 4 and 5) is in the voting stage and the remainder are under development. The inclusion of an eighth layer, the User Layer, has not been generally agreed. The IEC Fieldbus standard specifies only three of the seven layers of the ISO communication model. Pimentel (1989) discusses the important question: how important is it for fieldbus networks to be OSI consistent? As currently specified the fieldbus has a reduced architecture and it is therefore not fully consistent with the OSI model. Only the Echelon LonWorks implements the full seven-layer model. Pimentel notes that the Physical, Data Link and Application Layers are essential but the presentation and session layers are not needed. A four-layer architecture including the network layer is attractive because internetworking is facilitated. However, greater functionality could be achieved by replacing the network layer by the transport layer since network layer functionality can be supplied by bridges and routers. Even when a layer is removed some of its functionality must be incorporated in the adjacent layers. The reduced-layer models were introduced to improve the response time of fieldbus systems but it is clear that this can only be obtained by compromising the design philosophy of the open system interconnect.

The user layer, the user interface to the Application Layer, is receiving attention. Function block and data description language techniques have been developed by the process control industry. Early work concentrated on describing Application Services using the manufacturing message specification (MMS) terminology of MAP. An EIA Committee was responsible for the initial work on MMS. It was labelled RS511 by the committee but it is now more commonly referred to as MMS. It is a connectionless protocol which enables frames of data to be assembled into complete messages and it defines the message notation (i.e. it defines the purpose, length and value of each element in a message). MMS is now an international standard; ISO 9506.

The groups working on the Data Link Layer specification have experienced great difficulty in arriving at a document expressing the consensus view. One

A.13 IEC FIELDBUS (IEC DIS 1158)

group supports the use of centralized media access which ensures that sampling periods are maintained (this is deemed to be essential for good control loop performance) while another group supports the use of token passing media access control based on IEEE 802. Token passing allows changes to be easily made (because each token holder can manage its own scanning program) but sample rate jitter can be experienced at high traffic loads. Timed token methods can be envisaged with bus idle times sufficient to ensure that fixed scan times are achieved. A compromise document based on these methods has been produced.

The main features of the IEC electrical physical layer specification are the use of multidrop connections for up to 32 devices with twisted pair cables carrying asynchronous data transmission using Manchester coding and half-duplex communication. A voltage mode (with high- and low-speed options: 2.5 Mb/s, 1 Mb/s and 31.25 kb/s) and a current mode (1 Mb/s) is specified. A clip-on transformer connection is specified for the current mode bus with a maximum cable length of 750 m. The current mode bus uses a 20 kHz current-fed sine wave to supply power to remote devices and the 1 MHz digital data signal is added to the sine waveform. With the eventual inclusion of fibre optic and radio links, it is clear that a large number of options will be a feature of this standard.

Appendix B
Standards Organizations

Airlines Electronic Engineering Committee (ARINC)
Aeronautical Radio Inc.
2551 Riva Road
Annapolis
Maryland 21401
USA

American National Standards Institute (ANSI)
1430 Broadway
New York
NY 10018
USA

ARCNET Trade Association
3356 N. Arlington Hts. Road
Suite J
Arlington Heights
IL 60004
USA

Association Française de Normalisation (AFNOR)
Tour Europe - Cedex 7
92080 Paris La Defense
FRANCE

British Standards Institution
2 Park Street
London
W1A 2BS

Deutsche Institut für Normun (DIN)
Burggrafenstrasse 6
Postscah 1107
D-1000 Berlin
GERMANY

Electronics Industries Association (EIA)
Standards Sales
2001 Eye Street
Washington
DC 20006
USA

Institute of Electrical and Electronic Engineers (IEEE)
Standard Sales — IEEE Service Centre
445 Hoes Lane
Piscataway
NS 08854
USA

International Electrotechnical Commission (IEC)
1 Rue du Varembe
Case Postale 56
CH-1211 Geneva 20
SWITZERLAND

International Organisation for Standardization (ISO)
1 Rue du Varembe
Case Postale 56
CH-1211 Geneva 20
SWITZERLAND

Instrument Society of America (ISA)
Standards and Practices Working Group 50 (SP50)
PO Box 12277
Research Triangle Park
NC 27707
USA

Agency of Industrial Trade and Industry
Ministry of International Trade and Industry
1-3-1 Kasumigaseki, Chiyoda-Ku
Tokyo 100
JAPAN

National Electrical Manufacturers Association (NEMA)
Field Bus Working Group
2101L Street, N.W. Suite 300
Washington, DC 20037
USA

Appendix C
Integrated Circuits

C.1 ARCNET — COM20051*

Standard Microsystems Corporation
Component Products Division
80 Arkay Drive
Hauppauge
NY 11788
USA

CM20051 — INTEGRATED MICROCONTROLLER AND NETWORK INTERFACE

Features

- high performance/low cost
- microcontroller based on popular 8051 architecture
- intel compatible
- drop-in replacement for 80C32 PLCC
- network supports up to 255 nodes
- powerful network diagnostics
- maximum 512 byte packets
- duplicate node ID detection
- self-configuring network protocol
- retains all 8051 peripherals including serial I/O and two timers
- utilizes ARCNET token bus network engine
- requires no special emulators
- 5 Mbps to 156 kbps network data rate

---

* Much of this information is reproduced from the COM20051 data handbook 1993 by permission of Standard Microsystems Corporation

- network interface supports RS-485, twisted pair, coaxial, and fibre optic interfaces
- receive all mode allows any packet to be received

GENERAL DESCRIPTION

The COM20051 is a low-cost highly integrated microcontroller incorporating a high-performance network controller based on the ARCNET Token Bus Standard (ANSI 878.1). The COM20051 is based around the popular Intel 8051 architecture. The device is implemented using a microcontroller core compatible with the Intel 80C32 ROMless version of the 8051 architecture. The COM20051 is ideal for distributed control networking applications such as those found in industrial/machine controls, building/factory automation, consumer products, instrumentation and automobiles.

The COM20051 contains many features that are beneficial to embedded control applications. The microcontroller is a fully functional 16 MHz 80C32 that is compatible with the Intel 80C32. In contrast with other embedded controller/networking solutions, the COM20051 adds a fully featured, robust, powerful, and simple network interface while retaining all of the basic 8051 peripherals, such as the serial port and counter/timers.

In addition, the COM20051 supports an emulation mode that permits the use of a standard 80C32 emulator in conjunction with the COM20051 to develop software drivers for the network core. This mode is achieved by mapping the ARCNET network core into a 256 byte page of the external data memory space of the 80C32 instead of the SFR area, which would require a costly adapter for the emulator.

The networking core is based around an ARCNET Token Bus protocol engine that provides highly reliable and fault-tolerant message delivery at data rates ranging from 5 Mbps down to 156 kbps with message sizes varying from 1 to 508 bytes. The ARCNET protocol offers a simple, standardized and easily understood networking solution for any application. The network interface supports several media interfaces, including RS-485, coaxial and twisted pair in either bus or star topologies. The network interface incorporates powerful diagnostic features for network management and fault isolation. These include duplicate node ID detection, reconfiguration detection, receive all (monitor) mode, receiver activity, and token detection.

The COM20051 is essentially a network board-in-a-chip. It takes an 80C32 microcontroller core and an ARCNET controller and integrates them into a single device. ARCNET is a token passing-based protocol that combines powerful flow control, error detection, and diagnostic capabilities to deliver fast and reliable messages. The COM20051 supports a variety of data rates

C.I ARCNET – COM20051

(5 Mbps to 156 kbps), topologies (bus, star, tree), and media types (RS-485, coaxial, twisted pair, fibre optic, and powerline) to suit any type of application.

The ARCNET network core of the COM20051 contains many features that make network development simple and easy to comprehend. Diagnostic features, such as Receive All, Duplicate ID Detection, Reconfiguration Detection, Token and Receiver Detection, all combine to make the COM20051 simple to use and to implement in any environment. The ARCNET protocol itself is relatively simple to understand and very flexible. A wide variety of support products are available to assist in network development, such as software drivers, line drivers, boards, and development kits. The COM20051 implements a full-featured 16 MHz, Intel-compatible 80C32 microcontroller with all of the standard peripheral functions, including a full duplex serial port, two timer/counters, one 8 bit general purpose digital I/O port, and interrupt controller. The 8051 architecture has long been a standard in the embedded control industry for low-level data acquisition and control. ARCNET and the 8051 form a simple solution for many of today's and tomorrow's low-level networking solutions.

In addition to the 80C32 and the ARCNET network core, the COM20051 contains all the address decoding and interrupt routing logic to interface the network core to the 80C32 core. The integrated 8051/ARCNET combination provides an extremely cost-effective and space-efficient solution for industrial networking applications. The COM20051 can be used in a stand-alone embedded application, executing control algorithms or performing data acquisition and communicating data in a mater/slave or peer/peer configuration, or used as a slave processor handling communication tasks in a multiprocessing system.

Basic architecture

The COM20051 consists of four functional blocks: the 80C32 microcontroller core, ARCNET network cell (includes 1 K of buffer RAM), programmable address decoder, and programmable interrupt router. The internal architecture of the COM20051 is shown in Figure C.1.

The 80C32 microcontroller is a full ROMless implementation of the popular Intel 8051 series. The ARCNET network core is similar in architecture to SMC's popular COM20020 family of ARCNET controllers and retains the same command and status flags of previous ARCNET controllers. The programmable address decoder maps the ARCNET registers into a 256 byte page anywhere within the External Data Memory space of the 80C32. The ARCNET core was mapped to the External Data Memory space to simplify software and application

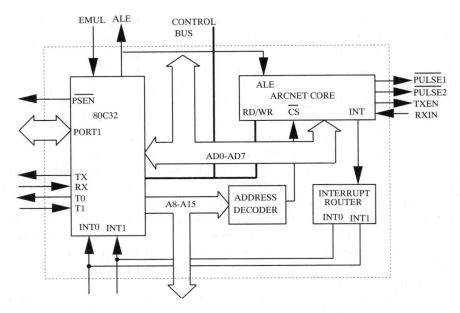

Figure C.1
Internal architecture of COM20051

development and for production test purposes. Access to the ARCNET core during software development is achieved by invoking the emulate mode. When the COM20051 is put into emulate mode, the internal microcontroller is put into a high impedance state, thus allowing an external in-circuit emulator (ICE) to program the ARCNET core. The advantage of this approach versus mapping the ARCNET registers into the internal memory (special function) area of the 80C32 is that dedicated software development tools will not be necessary to debug application software. Since a majority of 8051 applications use only a small portion of the Data Memory space, there is no penalty paid for use of the address space. There will also be no penalty in execution time, since cycle times for external data memory accesses and internal direct memory moves are identical. The network interrupt can be routed to either of the two external interrupt ports or can be assigned to one of the general purpose I/O ports. The ARCNET interrupt is internally wire ORed with the external interrupt pin to allow greater system flexibility.

The 80C32 microcontroller core is identical to the 16 MHz Intel 80C32 in all respects. Refer to the Intel Embedded Microcontrollers and Processors Databook, Volume 1, for details regarding the 8051 architecture, peripherals, instruction set and programming guide.

C.1 ARCNET – COM20051

The following differences apply to the COM20051:

- oscillator frequency is 40 MHz instead of 16 MHz. This is necessary to derive a 20 MHz clock for the ARCNET core; the processor still operates at 16 MHz
- EA/VPP pin — this pin must be tied to ground for normal internal processor operation; when tied to VCC, the COM20051 will enter the emulate mode
- unused pins — the COM20051 is packaged in a 44 pin PLCC; network I/O is generated on the four unused pins of the standard 80C32 PLCC package; no DIP package is available
- power down operation — the power down mode can only be used in conjunction when the internal oscillator is being used; if an external oscillator is used and the power down mode is invoked, damage may result to the oscillator and to the COM20051.

The COM20051 processor operates at 16 MHz and the network controller at a maximum 40 MHz clock rate. A single crystal oscillator is used to supply the two clocks: a 16 MHz processor clock and a 20 MHz network clock for the nominal 2.5 Mbps data rate. Pins 20 and 21 are designated as crystal inputs. When clocking with an external oscillator, pin 21 (XTAL1) functions as the clock input.

The COM20051 contains a unique feature called the emulate mode that most 8051-based peripheral devices do not accommodate. The emulate mode permits developers to access and program the internal ARCNET core using a standard low-cost 8032 emulator. This feature eliminates the need for expensive dedicated development equipment needed for other types of 8051-based peripheral devices.

Network protocol

Communication on the network is based on a token passing procol. Establishment of the network configuration and management of the network protocol are handled entirely by an internal microcoded sequencer. The 80C32 controller core transmits data by simply loading a data packet and its destination ID into the network core's RAM buffer, and issuing a command to enable the transmitter. When the ARCNET core next receives the token, it verifies that the receiving node is ready by first transmitting a FREE BUFFER ENQUIRY message. If the receiving node transmits an ACKnowledge message, the data packet is transmitted followed by a 16 bit CRC. If the receiving node cannot accept the packet (typically its receiver is inhibited), it transmits a Negative AcKnowledge message and the transmitter passes the token. Once it has been established that the receiving node can accept the packet and transmission

Table C.1
Example: IDLE LINE time-out at 2.5 Mbps = 82 µs. IDLE LINE time-out for 156.2 kbps is 82 µs × 16 = 1.3 ms. For 5 Mbps operation, all time-outs are scaled DOWN by two

CLOCK PRESCALER	DATA RATE W/20 MHz CLOCK	TIME-OUT SCALING FACTOR (MULTIPLY BY)
÷8	2.5 Mbps	1
÷12	1.25 Mbps	2
÷32	6.25 kbps	4
÷64	312.5 kbps	8
÷128	156.25 kbps	16

is complete, the receiving node verifies the packet. If the packet is received successfully, the receiving node transmits an ACKnowledge message (or nothing if it is not received successfully) allowing the transmitter to set the appropriate status bits to indicate successful or unsuccessful delivery of the packet. An interrupt mask permits the ARCNET core to generate an interrupt to the processor when selected status bits become true.

The ARCNET core is capable of supporting data rates from 156.25 kbps to 5 Mbps. The following protocol description assumes a 2.5 Mbps data rate. For slower data rates, an internal clock divider scales down the clock frequency. Thus all time-out values are scaled up as shown in Table C.1.

Network reconfiguration

A significant advantage of the ARCNET is its ability to adapt to changes on the network. Whenever a new node is activated or deactivated, a network reconfiguration is performed. When a new ARCNET node is turned on (creating a new active node on the network) or if the COM20051 has not received an invitation to transmit for 840 mS, or if a software reset occurs, the ARCNET node causes a network reconfiguration by sending a reconfigure burst consisting of eight marks and one space repeated 765 times. The purpose of this burst is to terminate all activity on the network. Since this burst is longer than any other type of transmission, the burst will interfere with the next invitation to transmit, destroy the token and keep any other node from assuming control of the line. The ARCNET protocol flow is shown in Figure C.2.

When any ARCNET node senses an idle line for greater than 82 µS, which occurs only when the token is lost, each node starts an internal timeout equal to 146 µs times the quantity 255 minus its own ID. It also sets the internally

C.1 ARCNET – COM20051

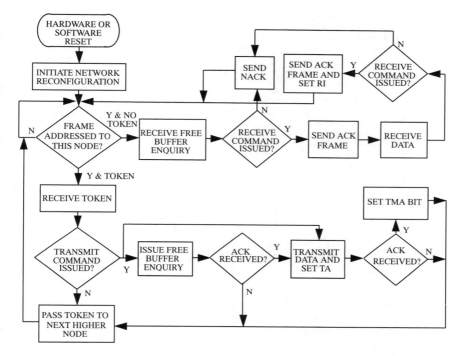

Figure C.2
Arcnet protocol flow

stored NID (next ID representing the next possible ID node) equal to its own ID. If the time-out expires with no line activity, the ARCNET core starts sending invitation to transmit with the Destination ID (DID) equal to the currently stored NID. Within a given network, only one node will time-out (the one with the highest ID number). After sending the invitation to transmit, the COM20051 waits for activity on the line. If there is no activity for 74.7 µS, the COM20051 increments the NID value and transmits another invitation to transmit using the NID equal to the DID. If activity appears before the 74.7 µS timeout expires, the COM20051 releases control of the line. During network reconfiguration, invitations to transmit are sent to all NIDs.

Each COM20051 on the network will finally have saved a NID value equal to the ID of the ARCNET node that it released control to. This is called the Next ID Value. At this point, control is passed directly from one node to the next with no wasted invitations to transmit being sent to IDs not on the network, until the next network reconfiguration occurs. When a node is powered off, the previous node attempts to pass the token to it by issuing an invitation to transmit. Since this node does not respond, the previous node times out and

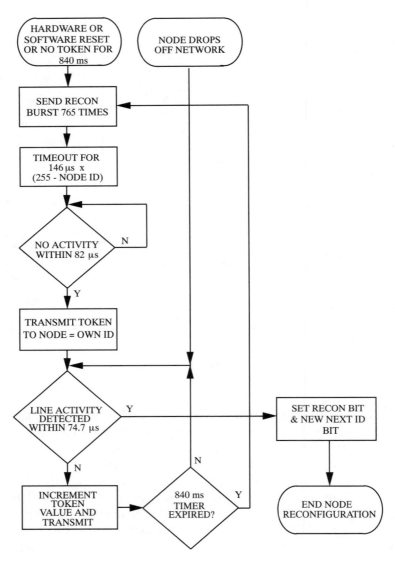

Figure C.3
ARCNET reconfiguration process

transmits another invitation to transmit to an incremented ID and eventually a response will be received. The network reconfiguration procedure is illustrated by the flow chart shown in Figure C.3.

The network reconfiguration time depends on the number of nodes in the network, the propagation delay between nodes, and the highest ID number on the network, but is typically within the range of 24–61 ms for 2.5 Mbps operation. There are three time-outs associated with the COM20051 operation. The values of these time-outs are controlled by bits 3 and 4 of the Configuration Register.

Broadcast messages

Broadcasting gives a particular node the ability to transmit a data packet to all nodes on the network simultaneously. ID zero is reserved for this feature and no node on the network can be assigned ID zero. To broadcast a message, the transmitting node's processor simply loads the RAM buffer with the data packet and sets the DID equal to zero.

Response time

The response time determines the maximum propagation delay allowed between any two nodes, and should be chosen to be larger than the round trip propagation delay between the two furthest nodes on the network plus the maximum turn around time (the time it takes a particular ARCNET node to start sending a message in response to a received message) which is approximately 12.7 μS. The round trip propagation delay is a function of the transmission media and network topology. For a typical system using RG62 coaxial cable in a baseband system, a one-way cable propagation delay of 31 μS translates to a distance of about 4 miles.

Idle time

The idle time is associated with a network reconfiguration. During a network reconfiguration one node will continually transmit invitations to transmit until it encounters an active node. All other nodes on the network must distinguish between this operation and an entirely idle line. During network reconfiguration, activity will appear on the line every 82 μS. This 82 μS is equal to the response time of 74.4 μS plus the time it takes the COM20051 to start retransmitting another message (usually another invitation to transmit).

Reconfiguration time

If any node does not receive the token within the reconfiguration time, the node will initiate a network reconfiguration. The ET2 and ET1 bits of the configuration register allow the network to operate over longer distances than the 4 miles stated earlier. The logic levels on these bits control the maximum distances over which the COM20051 can operate by controlling the three time-out values described above. For proper network operation, all nodes connected to the same network must have the same response time, idle time, and reconfiguration time.

Line protocol

The ARCNET line protocol is considered isochronous because each byte is preceded by a start interval and ended with a stop interval. Unlike asynchronous protocols, there is a constant amount of time separating each data byte. For a 2.5 Mbps data rate, each byte takes exactly 11 clock intervals of 400 ns each. As a result, one byte is transmitted every 4.4 μS and the time to transmit a message can be precisely determined. The line idles in a spacing (logic 0) condition. A logic 0 is defined as no line activity and a logic 1 is defined as a negative pulse of 200 ns duration. A transmission starts with an alert burst consisting of 6 unit intervals of mark (logic 1). Eight bit data characters are then sent, with each character preceded by 2 unit intervals of mark and one unit interval of space. Five types of transmission can be performed as described below:-

Invitations to transmit

An invitation to transmit is used to pass the token from one node to another and is sent by the following sequence:

- an alert burst
- an EOT (End Of Transmission: ASCII code 04H)
- two (repeated) DID (Destination ID) characters

ALERT BURST	EOT	DID	DID

Free buffer enquiries

A free buffer enquiry is used to ask another node if it is able to accept a packet of data. It is sent by the following sequence:

- an alert burst
- an ENQ (ENQuiry: ASCII node 85H)
- two (repeated) DID (Destination ID) characters

ALERT BURST	ENQ	DID	DID

Data packets

A data packet consists of the actual data being sent to another node. It is sent by the following sequence:

- an alert burst
- an SOH (Start Of Header-ASCII code 01H)
- an SID (Source ID) character
- two (repeated) ID (Destination ID) characters
- a single count character which is the 2s complement of the number of data bytes to follow if a short packet is sent, or OOH followed by a count character if a long packet is sent
- N data bytes where COUNT $= 256 - N$ (or $512 - N$ for a long packet)
- two CRC (Cyclic Redundancy Check) characters; the CRC polynomial used is: $X^{16} + X^{15} + X^2 + 1$.

ALERT BURST	SOH	SID	DID	DID	COUNT	data	data	CRC	CRC

Acknowledgements

An acknowledgement is used to acknowledge reception of a packet or as an affirmative response to free buffer enquiries and is sent by the following sequence:

- an alert burst
- an ACK (ACKnowledgement — ASCII code 86H) character

ALERT BURST	ACK

Negative acknowledgements

A negative acknowledgement is used as a negative response to free buffer enquiries and is sent by the following sequence:

- an alert burst
- a NAK (negative acknowledgement — ASCII code 15H) character

ALERT BURST	NAK

Microcontroller interface

All accesses to the internal RAM and the internal registers are controlled by the COM20051. The internal RAM is accessed via a pointer-based scheme and the internal registers are accessed via direct addressing. The ARCNET core bus interface is designed to be flexible so that it is independent of the 80C32 speed. The COM20051 provides for no wait state arbitration via direct addressing to its internal registers and a pointer based addressing scheme to access its internal RAM. The pointer may be used in auto-increment mode for typical sequential buffer emptying or loading, or it can be taken out of auto-increment mode to perform out of sequence accesses to the RAM. The data within the RAM is accessed through the data register. Data being read is prefetched from memory and placed into the data register for the microcontroller to read. During a write operation, the data are stored in the data register and then written into memory. Whenever the pointer is loaded for reads with a new value, data are immediately prefetched to prepare for the first read operation.

Transmission media interface

Traditional hybrid interface

The traditional hybrid interface (Figure C.4) is that which is used with previous ARCNET devices. The hybrid interface is recommended if the node is to be

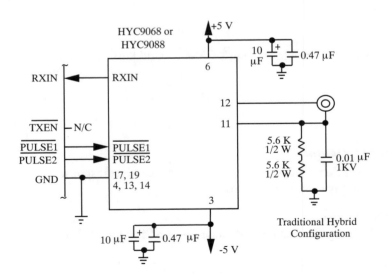

Figure C.4
DIPULSE hybrid configuration

C.I ARCNET – COM20051

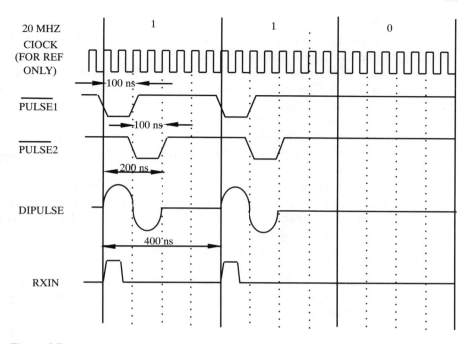

Figure C.5
DIPULSE waveform for data 1-1-0

placed in a network with other hybrid-interfaced nodes. The traditional hybrid interface is for use with nodes operating at 2.5 Mbps only. The transformer coupling of the hybrid offers isolation for the safety of the system and offers high common mode rejection. The traditional hybrid interface uses circuits like SMC's HYC9068 or HYC9088 to transfer the pulse-encoded data between the cable and the COM20051. The COM20051 transmits a logic 1 by generating two 100 nS non-overlapping negative pulses, PULSE1 and PULSE2. Lack of pulses indicates a logic 0. The PULSE1 and PULSE2 signals are sent to the Hybrid, which creates a 200 ns dipulse signal on the media. A logic 0 is transmitted by the absence of the dipulse. During reception, the 200 nS dipulse appearing on the media is coupled through the transformer of the LAN Driver, which produces a positive pulse at the RXIN pin of the COM20051. The pulse on the RXIN pin represents a logic 1. Lack of pulse represents a logic 0. Typically, RXIN pulses occur at multiples of 400 nS. The COM20051 can tolerate distortion (bit jitter) of plus or minus 100 nS and still correctly capture and convert the RXIN pulses to NRZ format. Figure C.5 illustrates the events which occur in transmission or reception of data consisting of 1, 1, 0.

Backplane configuration

The backplane configuration is recommended for cost-sensitive short-distance applications like backplanes and instrumentation. This mode is advantageous because it saves components, cost, and power. Since the backplane configuration encodes data differently than the traditional hybrid configuration, nodes utilizing the backplane configuration cannot communicate directly with nodes utilizing the traditional hybrid configuration.

The backplane configuration does not isolate the node from the media nor protect it from common mode noise, but common mode noise is less of a problem in short distances.

The COM20051 supplies a programmable output driver for backplane mode operation. A push/pull or open drain driver can be selected by programming the P1MODE bit of the setup register (see register descriptions for details.) The COM20051 defaults to an open drain output.

The backplane configuration provides for direct connection between the COM20051 and the media. Only one pull-up resistor (for open drain only) is required somewhere on the media (not on each individual node). The PULSE1 signal, in this mode, is an open drain or push/pull driver and is used to directly drive the media. It issues a 200 nS negative pulse to transmit a logic 1. Note that when used in the open-drain mode, the COM20051 does not have a fail/safe input on the RXIN pin.

The PULSE1 signal actually contains a weak pull-up resistor. This pull-up should not take the place of the resistor required on the media for open drain mode. In typical applications, the serial backplane is terminated at both ends and a bias is provided by the external pull-up resistor.

The RXIN signal is directly connected to the cable via an internal Schmitt trigger. A negative pulse on this input indicates a logic 1. Lack of pulse indicates a logic 0. For typical single-ended backplane applications, RXIN is connected to PULSE1 to make the serial backplane data line. A ground line (from the coaxial or twisted pair) should run in parallel with the signal. For applications requiring different treatment of the receive signal (like filtering or squelching), PULSE1 and RXIN remain as independent pins. External differential drivers/receivers for increased range and common mode noise rejection, for example, would require the signals to be independent of one another. When the device is in backplane mode, the clock provided by the PULSE2 signal may be used for encoding the data into a different encoding scheme or other synchronous operations needed on the serial data stream.

Differential driver configuration

The differential driver configuration is a special case of the backplane mode. It is a DC-coupled configuration recommended for applications like car-area

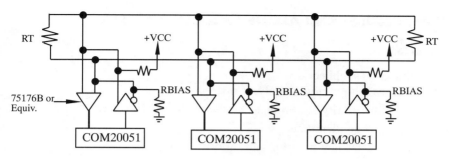

Figure C.6
COM20051 network using RS-485 differential transceivers

networks or other cost-sensitive applications which do not require direct compatibility with existing ARCNET nodes and do not require isolation.

The differential driver configuration cannot communicate directly with nodes utilizing the traditional hybrid configuration. Like the backplane configuration, the differential driver configuration does not isolate the node from the media.

The differential driver interface includes a RS485 driver/receiver to transfer the data between the cable and the COM20051 (as shown in Figure C.6). The PULSE1 signal transmits the data, provided the transmit enable signal is active. The PULSE1 signal issues a 200 nS negative pulse to transmit a logic 1. The RXIN signal receives the data. A negative pulse on this input indicates a logic 1. The RXIN signal receives the data. A negative pulse on this input indicates a logic 1. Lack of pulse indicates a logic 0. The transmitter portion of the COM20051 is disabled during reset and the PULSE1, PULSE2 and TXEN pins are inactive.

C.2 CAN – PCA82C200*

Philips Semiconductors Ltd
276 Bath Road
Hayes
Middlesex
UB3 5BX
UK

---

* This section reproduced by permission of Philips Semiconductors Ltd

PCA82C200: STAND-ALONE CAN CONTROLLER

Features

- multimaster architecture
- interfaces with a large variety of microcontrollers
- bus access priority (determined by the message identifier)
- 2032 message identifiers
- guaranteed latency time for high priority messages
- powerful error handling capability
- data length from 0 to 8 bytes
- configurable bus interface
- programmable clock output
- multicast and broadcast message facility
- non-destructive bit-wise arbitration
- non-return-to-zero (NRZ) coding/decoding with bit-stuffing
- programmable transfer rate (up to 1 Mbit/s)
- programmable output driver configuration
- suitable for use in a wide range of networks including the SAE networks Class A, B and C.

General description

The PCA82C200 is a highly integrated stand-alone controller for the controller area network (CAN) used within automotive and general industrial environments. The PCA82C200 contains all necessary features required to implement a high performance communication protocol. The PCA82C200 with a simple bus line connection performs all the functions of the physical and data-link layers. The application layer of an electronic control unit (ECU) is provided by a microcontroller, to which the PCA82C200 provides a versatile interface.

Functional description

The PCA82C200 contains all necessary hardware for a high performance serial network communication (see Figure C.7). The PCA82C200 controls the communication flow through the area network using the CAN-protocol. The PCA82C200 meets the following automotive requirements:

- short message length
- guaranteed latency time for urgent messages

C.2 CAN – PCA82C200

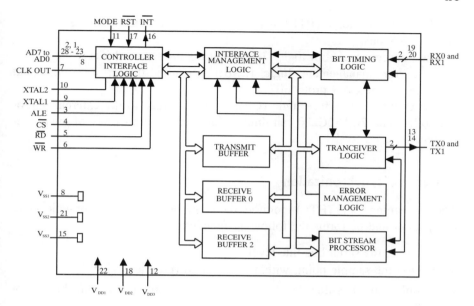

Figure C.7
Block diagram

- bus access priority, determined by the message identifier
- powerful error handling capability
- configuration flexibility to allow area network expansion.

The latency time defines the period between the initiation (transmission request) and the start of the transmission on the bus. Latency time is dependent on a variety of bus related conditions. In the case of a message being transmitted on the bus and one distortion on a variety of bus related conditions. In the case of a message being transmitted on the bus and one distortion the latency time can be up to 149 bit times (worst case).

The interface management logic interprets commands from the microcontroller, allocates the message buffers (TBF, RBF0 and RBF1) and provides interrupts and status information to the microcontroller.

The transmit buffer is a 10 byte memory into which the microcontroller writes messages which are to be transmitted over the CAN network.

The receive buffers RBF0 and RBF1 are each 10 byte memories which are alternatively used to store messages received from the CAN network. The CPU can process one message while another is being received.

The bit stream processor (BSP) is a sequencer, controlling the data stream between the transmit buffer, receive buffers (parallel data) and the CAN bus (serial data).

The bit timing logic (BTL) synchronizes the PCA82C200 to the bitstream on the CAN bus.

The transceiver control logic (TCL) controls the output driver.

The error management logic (EML) performs the error confinement according to the CAN protocol.

The controller interface logic (CIL) is the interface to the external microcontroller. The PCA82C200 can directly interface with a variety of microcontrollers.

Bus timing/synchronization

The bus timing logic (BTL) monitors the serial bus-line via the on-chip input comparator and performs the following functions.

- monitors the serial bus-line level
- adjusts the sample point, within a bit period (programmable)
- samples the bus-line level using majority logic (programmable, 1 or 3 samples)
- synchronization to the bit stream:
  — hard synchronization at the start of a message
  — resynchronization during transfer of a message.

The configuration of the BTL is performed during the initialization of the PCA82C200. The BTL uses the following three registers:

- control register (synch)
- bus timing register 0
- bus timing register 1.

A bit period is built up from a number of system clock cycles (t_{SCL}). One bit period is the result of the addition of the programmable segments TSEG1 and TSEG2 and the general segment SYNCSEG (see Figure C.8).

The incoming edge of a bit is expected during the SYNSEG state; this state corresponds to one system clock cycle ($1 \times t_{SCL}$).

Time segment 1 (TSEG1)

This segment determines the location of the sampling point within a bit period, which is at the end of TSEG1. TSEG1 is programmable from 1 to 16 system clock cycles.

The correction location of the sample point is essential for the correct functioning of a transmission. The following points must be taken into consideration.

C.2 CAN – PCA82C200

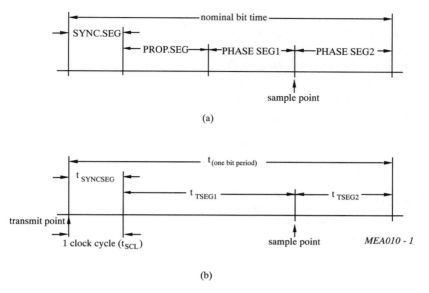

Figure C.8
(a) Bit period as defined by the CAN protocol; (b) bit period as implemented in the PCA82C200

- A start-of-frame causes all PCA82C200s to perform a 'hard synchronization' on the first recessive-to-dominant edge. During arbitration, however, several PCA82C200s may simultaneously transmit. Therefore it may require twice the sum of bus-line, input comparator and the output driver delay times until the bus is stable. This is the propagation delay which must be taken into consideration.
- To avoid sampling at an incorrect position, it is necessary to include an additional synchronization buffer on both sides of the sample point. The main reasons for incorrect sampling are:
  — incorrect synchronization due to spikes on the bus-line
  — slight variations in the oscillator frequency of each PCA82C200 in the network, which results in a phase error.

Time segment 2 (TSEG1)

This segment provides:

- additional time at the sample point for calculation of the subsequent bit levels (e.g. arbitration)
- synchronization buffer at right side of the sample point.

TSEG2 is programmable from 1 to 8 system clock cycles.

Synchronization jump width (SJW) defines the maximum number of clock cycles (t_{SCL}) a bit period may be reduced or increased by one resynchronization. SJW is programmable from 1 to 4 system clock cycles.

The propagation delay time (t_α) is calculated by summing the maximum propagation delay times of the physical bus, the input comparator and the output driver. The resulting sum is multiplied by 2 and then rounded up to the nearest multiple of t_{SCL}.

$$t_\alpha = 2 \times (\text{physical bus delay} + \text{input comparator delay} + \text{output driver delay})$$

Bit timing restrictions

Restrictions on the configuration of the bit timing are based on internal processing. The restrictions are:

- $t_{SEG2} \geq 2t_{SCL}$
- $t_{SEG2} \geq t_{SJW}$
- $t_{SEG1} \geq t_{SEG2}$
- $t_{SEG1} \geq t_{SJW} + t_\alpha$

The three sample mode (SAM = 1) has the effect of introducing a delay of one system clock cycle on the bus-line. This must be taken into account for the correct calculation of TSEG1 and TSEG2:

- $t_{SEG1} \geq t_{SJW} + t_\alpha + t_{SCL}$
- $t_{SEG2} \geq 3t_{SCL}$

Synchronization

Synchronization is performed by a state machine which compares the incoming edge with its actual bit timing and adapts the bit timing by hard synchronization or resynchronization.

Hard synchronization occurs only at the beginning of a message. The PCA82C200 synchronizes on the first incoming recessive-to-dominant edge of a message (being the leading edge of a message's start-of-frame bit).

Resynchronization occurs during the transmission of a message's bit stream to compensate for:

- variations in individual PCA82C200 oscillator frequencies
- changes introduced by switching from one transmitter to another (e.g. during arbitration).

As a result of resynchronization, either t_{TSEG1} may be increased by up to a maximum of t_{SJW} or t_{SEG2} may be decreased by up to a maximum of t_{SJW}:

C.2 CAN – PCA82C200

- $t_{SEG1} \geq t_{SCL}((\text{TSEG1} + 1) + (\text{SJW} + 1))$
- $t_{SEG2} \geq t_{SCL}((\text{TSEG2} + 1) - (\text{SJW} + 1))$

Note: TSEG1, TSEG2 and SJW are the programmed numerical values.

The phase error (e) of an edge is given by the position of the edge relative to SYNCSEG, measured in system clock cycles (t_{SCL}). The value of the phase error is defined as:

- $e = 0$, if the edge occurs within SYNCSEG
- $e > 0$, if the edge occurs within TSEG1
- $e < 0$, if the edge occurs within TSEG2.

The effect of resynchronization is:

- the same as that of a hard synchronization, if the magnitude of the phase error (e) is less or equal to the programmed value of t_{SJW}
- to increase a bit period by the amount of t_{SJW}, if the phase error is positive and the magnitude of the phase error is larger than t_{SJW}
- to decrease a bit period by the amount of t_{SJW}, if the phase error is negative and the magnitude of the phase error is larger than t_{SJW}.

Synchronization rules

The synchronization rules are as follows:

- only one synchronization within one bit time is used
- an edge is used for synchronization only if the value detected at the previous sample point differs from the bus value immediately after the edge
- hard synchronization is performed whenever there is a recessive-to-dominant edge during bus-idle
- all other edges (recessive-to-dominant and optionally dominant-to-recessive edges if the Synch bit is set HIGH, which are candidates for resynchronization) will be used with the following exception: a transmitting PCA82C200 will not perform a resynchronization as a result of a recessive-to-dominant edge with positive phase error, if only these edges are used for resynchronization; this ensures that the delay times of the output driver and input comparator do not cause a permanent in the bit time.

Communication protocol

The PCA82C200 bus controller supports the four different CAN protocol frame types for communication:

- data frame, to transfer data

- remote frame, request for data
- error frame, globally signal a (locally) detected error condition
- overload frame, to extend delay time of subsequent frames.

There are two logical bit representations used in the CAN protocol:

- a recessive bit on the bus-line appears only if all connected PCA82C200s send a recessive bit at that moment
- dominant bits always overwrite recessive bits, i.e. the resulting bit level on the bus-line is dominant.

Data frame

A data frame (Figure C.9) carries data from a transmitting PCA82C200 to one or more receiving PCA82C200s. A data frame is composed of seven different bit-fields:

- start-of-frame
- arbitration field
- control field
- data field (may have a length of zero)
- CRC field
- acknowledge field
- end-of-frame.

The start-of-frame bit signals the start of a data frame or remote frame. It consists of a single dominant bit used for hard synchronization of a PCA82C200 in receive mode.

The arbitration field consists of the message identifier and the RTR bit. In the event of simultaneous message transmissions by two or more PCA82C200s

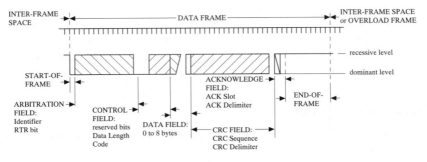

Figure C.9
Data frame

C.2 CAN – PCA82C200

the bus access conflict is solved by bit-wise arbitration, which is active during the transmission of the arbitration field.

The identifier is an 11 bit field used to provide information about the message, as well as the bus priority. It is transmitted in the order ID.10 to ID.0 (LSB). The situation that the seven most significant bits (ID.10 to ID.4) are all recessive must not occur.

An identifier does not define which particular PCA82C200 will receive the frame, as a CAN-based communication network does not discriminate between a point-to-point, multicast or broadcast communication.

A PCA82C200, acting as a receiver for certain information may initiate the transmission of the respective data by transmitting a remote frame to the network, addressing the data source via the identifier and setting the remote transmission request (RTR) bit HIGH (remote recessive bus level). If the data source simultaneously transmits a data frame containing the requested data, it uses the same identifier. No bus access conflict occurs due to the RTR bit being set LOW (data dominant bus level) in the data frame.

The control field consists of six bits. It includes two reserved bits (for future expansions of the CAN-protocol), transmitted with a dominant bus level, and is followed by the data length code (4 bits). The number of bytes in the (destuffed; number of data bytes to be transmitted/received) data field is indicated by the data length code. Admissible values of the data length code and hence the number of bytes in the (destuffed) data field, are 0 to 8. A logic 0 (logic 1) in the data length code is transmitted as a dominant (recessive) bus level, respectively.

The data, stored within the data field of the transmit buffer, are transmitted according to the data length code. Conversely, data of a received data frame will be stored in the data field of a receive buffer. Data is stored byte-wise both for transmission by the microcontroller and on reception by the PCA82C200. The most significant bit of the first data byte (lowest address) is transmitted/received first.

The cyclic redundancy field (CRC) consists of the CRC sequence (15 bits) and the CRC delimiter (1 recessive bit). The cyclic redundancy code (CRC) encloses the destuffed bit stream of the start-of-frame, arbitration field, control field, data field and CRC sequence. The most significant bit of the CRC sequence is transmitted/received first. This frame check sequence, implemented in the PCA82C200, is derived from a cyclic redundancy code best suited for frames with a total bit count of less than 127 bits. With start-of-frame (dominant bit) included in the code word, any rotation of the code word can be detected by the absence of the CRC delimiter (recessive bit).

The acknowledge field (ACK) consists of two bits, the acknowledge slot and the acknowledge delimiter, which are transmitted with a recessive level by the transmitter of the data frame. All PCA82C200s with received the matching

CRC sequence, report this by overwriting the transmitter's recessive bit in the acknowledge slot with a dominant bit. Thereby a transmitter, still monitoring the bus level recognizes that at least one receiver within the network has received a complete and correct message (i.e. no error was found). The acknowledge delimiter (recessive bit) is the second bit of the acknowledge field. As a result, the acknowledge slot is surrounded by two recessive bits; the CRC delimiter and the acknowledge delimiter.

All nodes within a CAN network may use all the information coming to the network by the PCA82C200s (shared memory concept). Therefore, acknowledgement and error handling are defined to provide all information in a consistent way throughout this shared memory. Hence, there is no reason to discriminate different receivers of a message in the acknowledge field. If a node is disconnected from the network due to bus failure, this particular node is no longer part of the shared memory. To identify a lost node additional and application specific precautions are required.

Each data frame or remote frame is delimited by the end-of-frame bit sequence which consists of seven recessive bits (exceeds the bit stuff width by two bits). Using this method a receiver detects the end of a frame independent of a previous transmission error because the receiver expects all bits up to the end of the CRC sequence to be coded by the method of bit-stuffing. The bit-stuffing logic is deactivated during the end-of-frame sequence.

Remote frame

A PCA32C200, acting as a receiver for certain information may initiate the transmission of the respective data by transmitting a remote frame to the network, addressing the data source via the identifier and setting the RTR bit HIGH (remote; recessive bus level). The remote frame is similar to the data frame with the following exceptions:

- RTR bit is set HIGH
- data length code is ignored
- no data field contained.

Note that the data length code value should be the same as for the corresponding data frame (although this is ignored for a remote frame).

A remote frame is composed of six different bit fields:

- start-of-frame
- arbitration field
- control field
- CRC-field
- acknowledge field
- end-of-frame.

C.2 CAN – PCA82C200

Error frame

The error frame consists of two different fields. The first field is accomplished by the superimposing of error flags contributed from different PCA82C200s. The second field is the error delimiter.

There are two forms of error flag:

- active error flag, consists of six consecutive dominant bits
- passive error flag, consists of six consecutive recessive bits unless it is overwritten by dominant bits from other PCA32C200s.

An error-active PCA82C200 detecting an error condition signals this by transmission of an active error flag. This error flag's form violates the bit-stuffing law applied to all fields, from start-of-frame to CRC delimiter, or destroys the fixed form of the fields acknowledge field or end-of-frame. Consequently, all other PCA82C200s detect an error condition and start transmission of an error flag. Therefore the sequence of dominant bits, which can be monitored on the bus, results from a superposition of different error flags transmitted by individual PCA82C200s. The total length of this sequence varies between six (min) and twelve (max) bits.

An error-passive PCA82C200 detecting an error condition tries to signal this by transmission of a passive error flag. The error-passive PCA82C200 waits for six consecutive bits with identical polarity, beginning at the start of the passive error flag. The passive error flag is complete when these six identical bits have been detected.

The error delimiter consists of eight recessive bits and has the same format as the overload delimiter. After transmission of an error flag, each PCA82C200 monitors the bus-line until it detects a transition from a dominant-to-recessive bit level. At this point in time, every PCA82C200 has finished sending its error flag and all PCA82C200s start transmission of seven recessive bits (plus the recessive bit at dominant-to-recessive transition, results in a total of eight recessive bits). After this event and an intermission field all error-active PCA82C200s within the network can start a transmission simultaneously.

If a detected error is signalled during transmission of a data frame or remote frame, the current message is spoiled and a retransmission of the message is initiated.

If a PCA82C200 monitors any deviation of the error frame, a new error frame will be transmitted. Several consecutive error frames may result in the PCA82C200 becoming error-passive and leaving the network unblocked.

In order to terminate an error flag correctly, an error-passive CAN bus controller requires the bus to be bus-idle for at least three bit periods (if there is a local error at an error-passive receiver). Therefore a CAN bus should not be 100% permanently loaded.

Overload frame

The overload frame consists of two fields, the overload flag and the overload delimiter. There are two conditions in the CAN-protocol which lead to the transmission of an overload flag:

- condition 1: receiver circuitry require more time to process the current data before receiving the next frame (receiver not ready)
- condition 2: detection of dominant bit during intermission field.

The transmission of an overload frame may only start:

- condition 1: during the first bit period of an expected intermission field
- condition 2: one bit period after detecting the dominant bit during intermission field.

The PCA82C200 will never initiate transmission of a condition 1 overload frame and will only react on a transmitted condition 2 overload frame, according to the CAN protocol. No more than two overload frames are generated to delay a data frame or a remote frame. Although the overall form of the overload frame corresponds to that of the error frame, an overload frames does not initiate or require the retransmission of the preceding frame.

The overload flag consists of six dominant bits and has a similar format to the error flag. The overload flags form corrupts the fixed form of the intermission field. All other PCA82C200s detecting the overload condition also transmit an overload flag.

The overload delimiter consists of eight recessive bits and takes the same form as the error delimiter. After transmission of an overload flag, each PCA82C200 monitors the bus-line until it detects a transition from a dominant-to-recessive bit level. At this point in time, every PCA82C200 has finished sending its overload flag and all PCA82C200s start simultaneously transmitting seven more recessive bits.

Inter-frame space

Data frames and remote frames are separated from preceding frames (all types) by an inter-frame space, consisting of an intermission field and a bus-idle. Error-passive PCA82C200s also send a suspend transmission after transmission of a message. Overload frames and error frames are not preceded by an inter-frame space.

The intermission field consists of three recessive bits. During an intermission period, no frame transmissions will be started by any PCA82C200. An intermission is required to have a fixed time period to allow a CAN-controller to execute internal processes prior to the next receive or transmit task.

C.2 CAN – PCA82C200

The bus-idle time may be of arbitrary length (min. 0 bit). The bus is recognized to be free and a CAN-controller having information to transmit may access the bus. The detection of a dominant bit level during bus-idle on the bus is interpreted as the start-of-frame.

Bus organization

Bus organization is based on five basic rules described in the following paragraphs.

Bus access: PCA82C200s only start transmission during the bus-idle state. All PCA82C200s synchronize on the leading edge of the start-of-frame (hard synchronization).

Arbitration: if two or more PCA82C200s simultaneously start transmitting, the bus access conflict is solved by a bit-wise arbitration process during transmission of the arbitration field.

During arbitration every transmitting PCA82C200 compares its transmitted bit level with the monitored bus level. Any PCA82C200 which transmits a recessive bit and monitors a dominant bus level immediately becomes the receiver of the higher priority message on the bus without corrupting any information on the bus. Each message contains a unique identifier and a RTR bit describing the type of data within the message. The identifier together with the RTR bit implicitly define the message's bus access priority. During arbitration the most significant bit of the identifier is transmitted first and the RTR bit last. The message with the lowest binary value of the identifier and RTR bit has the highest priority.

A data frame has higher priority than a remote frame due to its RTR bit having a dominant level.

For every data frame there is a unique transmitter. For reasons of compatibility with other CAN-bus controllers, use of the identifier bit pattern ID = $1111111XXXX_b$ (X being bits of arbitrary level) is forbidden. The number of available different identifiers is 2032 ($2^{11} - 2^4$).

Coding/decoding

The following bit fields are coded using the bit-stuffing technique: start-of-frame, arbitration field, control field, data field, CRC sequence.

When a transmitting PCA82C200 detects five consecutive bits of identical polarity to be transmitted, a complementary (stuff) bit is inserted into the transmitted bit-stream.

When a receiving PCA82C200 has monitored five consecutive bits with identical polarity in the received bit streams of the above described bit fields, it automatically deletes the next received (stuff) bit. The level of the deleted stuff bit has to be the complement of the previous bits; otherwise a stuff error will be detected and signalled.

The remaining bit fields or frames are of fixed form and are not coded or decoded by the method of bit-stuffing.

The bit-stream in a message is coded according to the non-return-to-zero (NRZ) method, i.e. during a bit period the bit level is held constant either recessive or dominant.

Error signalling: a PCA82C200 which detects an error condition, transmits an error flag. Whenever a bit error, stuff error, form error or an acknowledgement error is detected, transmission of an error flag is started at the next bit.

Whenever a CRC error is detected, transmission of an error flag starts at the bit following the acknowledge delimiter, unless an error flag for another error condition has already started. An error flag violates the bit-stuffing law or corrupts the fixed form bit fields. A violation of the bit-stuffing law affects any PCA82C200 which detects the error condition. These devices will also transmit an error flag.

An error-passive PCA82C200 which detects an error condition, transmits a passive error flag. A passive error flag is not able to interrupt a current message at different PCA82C200s, but this type of error flag may be ignored by other PCA82C200s. After having detected an error condition, an error-passive PCA82C200 will wait for six consecutive bits with identical polarity and when monitoring them, interpret them as an error flag.

After transmission of an error flag, each PCA82C200 monitors the bus-line until it detects a transition from a dominant-to-recessive bit level. At this point in time, every PCA82C200 has finished transmitting in its error flag and all PCA82C200s start transmitting seven additional recessive bits.

The message format of a data frame or remote frame is defined in such a way, that all detectable errors can be signalled within the message transmission time, and therefore it is very simple for a PCA82C200 to associate an error frame to the corresponding message and to initiate retransmission of the corrupted message.

If a PCA82C200 monitors any deviation of the fixed form of an error frame, it transmits a new error frame.

Overload signalling: some CAN controllers (but not the PCA82C200) require to delay the transmission of the next data frame or remote frame by transmitting one or more overload frames. The transmission of an overload frame must start during the first bit of an expected intermission. Transmission of overload frames which are reactions on a dominant bit during an expected intermission field, start one bit after this event.

Though the format of overload frame and error frame are identical, they are treated differently. Transmission of an overload frame during intermission

C.2 CAN – PCA82C200

field does not initiate the retransmission of any previous data frame or remote frame.

If a CAN controller which transmitted an overload frame monitors any deviation of its fixed form, it transmits an error frame.

Error detection

The processes described in the following paragraphs are implemented in the PCA82C200 for error detection.

Bit error: a transmitting PCA82C200 monitors the bus on a bit-by-bit basis. If the bit level monitored is different from the transmitted one, a bit error is signalled. The exceptions are as follows. During the arbitration field, a recessive bit can be overwritten by a dominant bit. In this case, the PCA82C200 interprets this as a loss of arbitration. During the acknowledge slot, only the receiving PCA82C200s are able to recognize a bit error.

Stuff error: the following bit fields are coded using the bit-stuffing technique: start-of-frame, arbitration field, control field, data field, CRC sequence.

There are two possible ways of generating a stuff error. The disturbance generates more than the allowed five consecutive bits with identical polarity. These errors are detected by all PCA82C200s. A disturbance falsifies one or more of the five bits preceding the stuff bit. This error situation is not recognized as a stuff error by the receivers. Therefore, other error detection processes may detect this error condition such as: CRC check, format violation at the receiving PCA82C200s or bit error detection by the transmitting PCA82C200.

CRC error: to ensure the validity of a transmitted message, all receivers perform a CRC check. Therefore, in addition to the (destuffed) information digits (start-of-frame up to data field), every message includes some control digits (CRC sequence; generated by the transmitting PCA82C200 of the respective message) used for error detection.

The code used for the PCA82C200 bus controller is a (shortened) BCH code, extended by a parity check, and has the following attributes:

- up to five single bit errors are 100% detected, even if they are distributed randomly within the code
- all single bit errors are detected if their total number (within the code) is odd
- the residual error probability of the CRC check amounts to 3×10^{-5}; as an error may be detected not only by the CRC check but also by other detection processes, the residual probability is several magnitudes less than 3×10^{-5} for undetected errors

- 127 bits as the maximum length of the code
- 112 bits as the maximum number of information digits (max. 83 bits are used by PCA82C200)
- length of the CRC sequence amounts to 15 bits
- Hamming distance $d = 6$.

As a result, $(d - 1)$ random errors are detectable (some exceptions exist).

The CRC sequence is calculated by the following procedure:

- the destuffed bit stream consisting of start-of-frame up to the data field (if present) is interpreted as a polynomial with coefficients of 0 or 1
- this polynomial is divided (modulo 2) by the following generator polynomial

$$f(X) = (X^{14} + X^9 + X^8 + X^6 + X^5 + X^4 + X^2 + X + 1)(X + 1)$$
$$= 1100010110011001_b$$

The remainder of this polynomial division is the CRC sequence which includes a parity check. Burst errors are detected up to a length of 15 (degree of $f(X)$). Multiple errors (number of disturbed bits at least $d = 6$) are not detected with a residual error probability of $2^{-15} (3 \times 10^{-5})$ by CRC check only.

Form error: form errors result from violation of the fixed form of the following bit fields: end-of-frame, intermission, acknowledge delimiter, CRC delimiter.

During the transmission of these bit fields an error condition is recognized if a dominant bit level instead of a recessive one is detected.

- acknowledgement error: this is detected by a transmitter whenever it does not monitor a dominant bit during the acknowledge slot
- error detection by an error flag of another PCA82C200: the detection of an error is signalled by transmitting an error flag; an active error flag causes a stuff error, a bit error or a form error at all other PCA82C200s.
- error detection capabilities: errors which occur at all PCA82C200s (global errors) are 100% detected; for local errors, i.e. for errors occurring at some PCA82C200s only, the shortened BCH code, extended by a parity check, has the following error detection capabilities.

Error confinement (definitions)

A PCA82C200 which has too many unsuccessful transmission, relative to the number of successful transmissions, will enter the bus-off state. It remains in this state, neither receiving nor transmitting messages until the reset request bit is set LOW (absent) and both error counters are set to 0.

A PCA82C200 which has received a valid message correctly, indicates this to the transmitter by transmitting a dominant bit level on the bus during the acknowledge slot, independent of accepting or rejecting the message.

An error-active PCA82C200 is in its normal operating state able to receive and to transmit normally and also to transmit an active error flag.

An error-passive PCA82C200 may transmit or receive messages normally. In the case of a detected error condition it transmits a passive error flag, instead of an active error flag. Hence the influence on bus activities by an error-passive PCA82C200 (e.g. due to a malfunction) is reduced.

After an error-passive PCA82C200 has transmitted a message, it sends eight recessive bits after the intermission field and then checks for bus-idle. If during suspend transmission another PCA82C200 starts transmitting a message the suspended PCA82C200 will become the receiver of this message; otherwise being in bus-idle it may start to transmit a further message.

A PCA82C200 which either was either switched off or is in the bus-off state, must run a start-up routine in order to:

- synchronize with other available PCA82C200s, before starting to transmit. Synchronizing is achieved, when 11 recessive bits, equivalent to acknowledge delimiter, end-of-frame and intermission field, have been detected (bus-free).
- wait for other PCA82C200s without passing into the bus-off state (due to a missing acknowledge), if there is no other PCA82C200 currently available.

Aims of error confinement

The microcontroller must be informed when there are long-lasting disturbances and when bus activities have returned to normal operation. During long-lasting disturbances, a PCA82C200 enters the bus-off state and the microcontroller may use default values.

Minor disturbances of bus activities will not affect a PCA82C200. In particular, a PCA82C200 does not enter the bus-off state or inform the microcontroller of a short lasting bus disturbance.

The rules for error confinement are defined by the CAN protocol specification (and implemented in the PCA82C200), in that the PCA82C200, being nearest to the error-locus, reacts with a high probability the quickest (i.e. becomes error-passive or bus-off). Hence errors can be localized and their influence on normal bus activities is minimized.

All PCA82C200s contain a transmit error counter and a receive error counter, which registers errors during the transmission and the reception of messages, respectively.

If a message is transmitted or received correctly, the count is decreased. In the event of an error, the count is increased. The error counters have an non-proportional method of counting: an error causes a larger counter increase than a correctly transmitted/received message causes the count to decrease. Over a period of time this may result in an increase in error counts, even if there are fewer corrupted messages than uncorrupted ones. The level of the error counters reflect the relative frequency of disturbances. The ratio of increase/decrease depends on the acceptable ratio of invalid/valid messages on the bus and is hardware-implemented to eight.

If one of the error counters exceeds the warning limit of 96 error points, indicating an appreciable accumulation of error conditions, this is signalled by the PCA82C200 (error status, error interrupt).

A PCA82C200 operates in the error-active mode until it exceeds 127 error points on one of its error counters. At this point it will enter the error-passive state.

A transmit error which exceeds 255 error points results in the PCA82C200 entering the bus-off state.

Philips Semiconductors and North American Philips Corporation Products are not designed for use in life support appliances, devices or systems where malfunction of a Philips Semiconductors and North American Philips Corporation Product can reasonably be expected to result in a personal injury. Philips Semiconductors and North American Philips Corporation customers using or selling Philips Semiconductors and North American Philips Corporation Products for use in such applications do so at their own risk and agree to fully indemnify Philips Semiconductors and North American Philips Corporation for any damages resulting from such improper use or sale.

C.3 ECHELON NEURON CHIP* (MOTOROLA MC143150/MC143120; TOSHIBA TMPN 3150/3120)

Echelon Europe Ltd
Elsinore House
77 Fulham Palace Road
London
W6 8JA
UK

---

* Much of this section is reproduced from the data handbook on the Echelon Neuron Chip by permission of the Echelon Corporation

NEURON CHIP DISTRIBUTED COMMUNICATIONS AND CONTROL PROCESSOR

The 3150 and 3120 Neuron chips are communications and control processors which enable the development of interoperable products. They provide systems designers with a wide range of features to accelerate product development for distributed sense and control applications. Services at every layer of the OSI networking model have been implemented in the included LonTalk firmware-based protocol and are easily and optionally invoked. In addition, common I/O routines are provided and can be accessed from ROM using the high-level Neuron C programming language.

The neuron chips are programmed with the LonBuilder Developer's Workbench which provides a sophisticated platform for software development and system integration over twisted pair, powerline, RF and other networking media. The Neuron 3150 chip addresses external memory to access the LonTalk protocol firmware, while the 32 pin Neuron 3120 chip includes a 10 k byte on-chip ROM containing this firmware.

- Three 8 bit pipelined processors for concurrent processing of application code and network packets.
- Eleven-pin I/O port programmable in 34 modes for fast application program development.
- Two 16 bit timer/counters for measuring and generating I/O device waveforms.
- Five-pin communications port that supports direct connect and network transceiver interfaces.
- 1024 or 2048 bytes of static RAM for buffering network data and storing network variables.
- 512 bytes (1024 bytes in the 3120E1 and 2048 bytes in the 3120E2) of EEPROM with on-chip charge pump for flexible storage of, for example, network addresses and communication parameters (and application code in the case of the 3120).
- External memory interface to support large application programs (Neuron 3150 only).
- Sleep mode to reduce power consumption.
- Unique 48 bit ID number in every device to facilitate network installation and management.
- 10 k bytes of preprogrammed ROM containing communication protocol and I/O firmware (Neuron 3120).

These integrated circuits contain firmware which has license restrictions. Sample Neurons can be obtained from Motorola and Toshiba after signing a developer's license agreement with the Echelon Corporation. Production procurement of the Neuron circuits can be acquired from Motorola and Toshiba only after signing an OEM license agreement with Echelon Corporation.

INTRODUCTION

The 3150 (Figure C.10) and 3120 Neuron chips are sophisticated VLSI devices that make it possible to implement low-cost control network applications. Through a unique combination of hardware and firmware, they provide all the key functions necessary to process inputs from sensors and control devices intelligently, and propagate control information across a variety of network media. Specifically, the 3150 and 3120 Neuron chips with the LonBuilder Developer's Workbench offer to the system engineer:

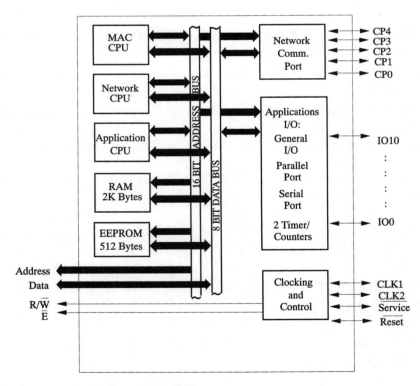

Figure C.10
Neuron 3150 chip block diagram

- easy implementation of distributed sense and control networks
- flexible reconfiguration capability after network installation
- management of LonTalk protocol messages on the network
- an object-oriented high level environment for system development.

The Neuron 3150 chip is designed for sense and control systems which require large application programs. An external memory interface allows the system designer to use 42 k of the available 64 k of address space for application program storage. The Neuron 3150 has no ROM on the chip. The communications protocol, operating system and 34 I/O function object code is provided by the Developer's Workbench. The protocol and application code can be located in external ROM, EEPROM, NVRAM or battery-backed static RAM or FLASH memory.

The Neuron 3120 chip has no external memory interface and is designed for applications which require smaller application programs. It contains 10 k of masked ROM that implements the communications protocol, operating system and the 34 I/O functions that can be assessed by the application program. The application program resides in the internal 512 bytes (or 1024 bytes on the 3120E1 or 2048 bytes on the 3120E2) of EEPROM and utilizes the masked ROM firmware that it needs for the specific application.

Both parts have an eleven-pin I/O interface with integrated hardware and software for connecting to motors, valves, display drivers, A/D convertors, pressure sensors, thermistors, switches, relays, triacs, tachometers, other microprocessors, modems, etc. They each have three processors, of which two interact with a communication subsystem to make the transfer of information from node to node in a distributed control system an automatic process.

LONWORKS OVERVIEW AND ARCHITECTURE

LonWorks is a complete platform to implement control network systems. LonWorks-based systems consist of intelligent devices or nodes that interact with their environment, and communicate with one another over a variety of communications media using a common, message-based control protocol.

LonWorks technology includes all of the elements required to design, deploy and support network systems, specifically the following components: Neuron 3150 chips and Neuron 3120 chips, LonTalk Protocol, LonWorks Transceivers, LonBuilder Developer's Workbench, LonManager network modules, hardware plus software, connectivity products, PC interfaces, and services such as training and customer support.

The Neuron chip is a VLSI component that performs the network and application-specific processing within a node. A node (Figure C.11) typically

APPENDIX C: INTEGRATED CIRCUITS

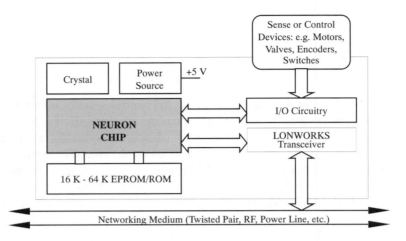

Figure C.11
Neuron 3150 chip in a typical node block diagram

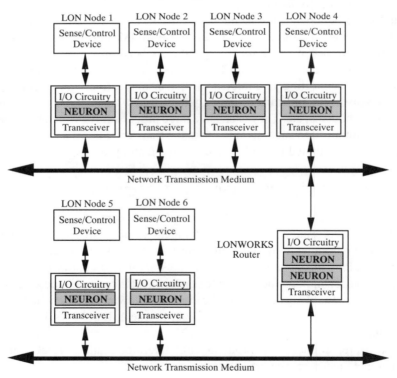

Figure C.12
Neuron chips in a LonWorks control network

consists of a Neuron chip, a power source, a transceiver for communicating over the network medium, and circuitry for interfacing to the device being controlled or monitored. The specific circuitry will depend on the networking medium and application. The Neuron chip in a LonWorks control network is shown in Figure C.12.

PROCESSING UNITS

The three identical 8 bit state machines form a symmetric multiprocessor, each CPU executing a different micro-cycle within a three-phase minor cycle at any one time. Each CPU has its own register set but all three CPUs share 8 bit data and address ALUs and memory access circuitry. This reduces die size without affecting performance.

The CPU architecture is stack-oriented; one 8 bit wide stack is used for data references, and the ALU operates on the TOS (top of stack) register and the next entry in the data stack which is in RAM. A second stack stores the return addresses for CALL instructions, and may also be used for temporary data storage. This zero-address architecture leads to very compact code.

Figure C.13 shows the layout of a base page, which may be up to 256 bytes long. Each of the three CPUs uses a different base page, whose address is given by the contents of the BP register of that CPU. The top of the data stack is in the 8 bit TOS register, and the next element in the data stack is at the location within the base page at the offset given by the contents of the SP register. The data stack grows from low memory towards high memory, and the return stack grows from high memory towards low memory.

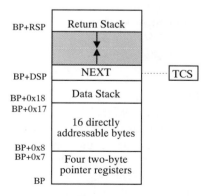

Figure C.13
Base page memory layout

MEMORY

The Neuron 3150 chip has 512 bytes of in-circuit programmable EEPROM. The 3120E1 has 1k EEPROM and the 3120E2 has 2k EEPROM with 2k RAM. All but eight bytes of the EEPROM can be written under program control using an on-chip charge pump to generate the required programming voltage. The remaining eight bytes are written during manufacture, and contain a unique 48 bit identifier for each part, plus 16 bits for the manufacturer's device code. Erase time and write time are each 20 ms per byte, and each byte of the EEPROM may be written up to 10 000 times with no data loss. For both Neuron chips the EEPROM stores the installation-specific information such as network addresses and communications parameters. For the Neuron 3120 chip, EEPROM memory also stores the application program generated by the Developer's Workbench. The application code for the Neuron 3150 chip may either be stored on-chip, or off-chip in external memory. Note that when the Neuron chip is not within the specified power supply voltage range, a pending or on-going EEPROM write is not guaranteed, and while there is built-in protection to prevent EEPROM corrupt on during power-down, it is important to hold the RESET pin low whenever V_{DD} is below its minimum operating level to avoid this possibility. Second generation 3150, 3120 parts have an integral low voltage indicator (LVI) to prevent this from occurring. There is also enhanced EEPROM failure recover in the firmware.

The Neuron 3150 chip has 2048 bytes of static RAM and the Neuron 3120E1 chip has 1024 bytes of static RAM (2k RAM in the 3120E2). The RAM state is retained as long as power is applied to the device, even in 'sleep' mode.

The Neuron 3120 chip contains 10 240 bytes of pre-programmed ROM. This memory contains the LonWorks firmware, including the LonTalk protocol code, real-time task scheduler and application function libraries. The Neuron 3150 chip accesses external memory for all of these. The object code is supplied with the LonBuilder development system.

INPUT/OUTPUT

Eleven bi-directional I/O pins are usable in several different configurations to provide flexible interfacing to external hardware and access to the internal timer/counters. The level of output pins may be read back by the CPU.

Pins I/O4–I/O7 have programmable pull-ups (current sources). They are enabled or disabled with a compiler directive. Pins I/O0–I/O3 have high current sink capability (20 ma at 0.8 V). The others have the standard sink capability (1.4 ma at 0.4 V). All pins (I/O0–I/O10) have TTL-level inputs with hysteresis.

C.3 ECHELON NEURON CHIP

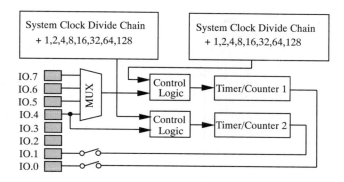

Figure C.14
Timer/counter circuits

Two 16 bit timer/counters are implemented as a load register writeable by the CPU, a 16 bit counter, and a latch readable by the CPU. The 16 bit registers are accessed a byte at a time. Both the Neuron 3150 chip and Neuron 3120 chip have one timer/counter whose input is selectable among pins I/O4-I/O7, and whose output is pin I/O0, and a second timer/counter with input from pin I/O4 and output to pin I/O1 (Figure C.14). Note that no I/O pins are dedicated to timer/counter functions. If for example, timer/counter 1 is used for input signals only, then I/O0 is available for other input or output functions. Timer/counter clock and enable inputs may be from external pins, or from scaled clocks derived from the system clock; the clock rates of the two timer/counters are independent of each other. External clock actions occur optionally on the rising edge, the falling edge, or both rising and falling edges of the input.

A minimum hardware configuration using the flash memory option is shown in Figure C.15. Note that the hardware design must ensure that both the Neuron Chip and memory timing specifications are met.

Because one of the internal CPUs is always using the address bus, sharing the address and data bus (and memory) with another MPU address and data

Table C.2
External memory interface pins

PIN DESIGNATION	DIRECTION	FUNCTION
A0-A15	Output	Address pins
D0-D7	Input/Output	Data pins
-E	Output	Enable clock
R/-W	Output	Read/not Write select

APPENDIX C: INTEGRATED CIRCUITS

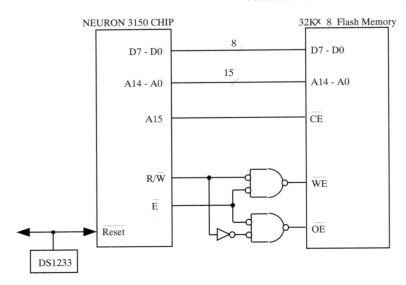

Figure C.15
Minimum Neuron 3150 chip memory interface

bus can only be accomplished with the use of external three state bus drivers on the 16 address lines. This method is not recommended or emulated with the LonBuilder development system. The preferred method of interfacing the Neuron chip to another MPU is through the 11 I/O pins. There are parallel functions and serial functions for this purpose which are easily implemented using the Neuron C programming language.

Table C.3
Communications port pin characteristics

Pin	Drive Current (mA)	Direct Mode Differential	Direct Mode Single-Ended	Special Purpose Mode
CP.0	1.4	RX + (in)	RX (in)	RX (in)
CP.1	1.4	RX − (in)	TX (out)	TX (out)
CP.2	35	TX + (out)	TX enable (out)	Bit Clock (out)
CP.3	35	TX − (out)	sleep (out)	Sleep (out) or Wake-up input
CP.4	1.4	CDet (in)	CDet (in)	Frame Clock (out)

RX - receiver
TX - transmitter
CDet - collision detect

NETWORK COMMUNICATIONS

The five-pin communications port is the network connection that interfaces the processor to a wide variety of media interfaces (network transceivers see Table C.3). The communications port may be configured to operate in one of two modes: Direct Mode (single ended or differential) or Special Purpose Mode. The internal transceiver block diagram is shown in Figure C.16.

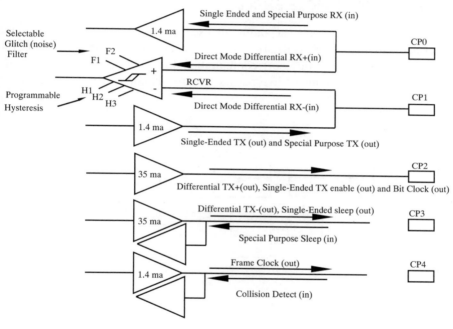

Figure C.16
Internal transceiver block diagram

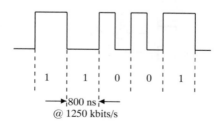

Figure C.17
Direct mode differential Manchester data encoding

In Direct mode, the processor communicates with the transceiver using a stream of Differential Manchester encoded data (see Figure C.17). This guarantees a transition in every bit period for purposes of synchronizing the receiver clock. Zero/one data is indicated by the presence/absence (respectively) of a second transition halfway between clock transitions. This mode is polarity-insensitive.

DIRECT MODE SINGLE ENDED

The single-ended mode is most commonly used with external active transceivers interfacing to media such as RF, IR, optical fibre, and coaxial cable. Figure C.18 shows the communications port configuration for the single-ended mode of operation. Data communication occurs via the single-ended (with respect

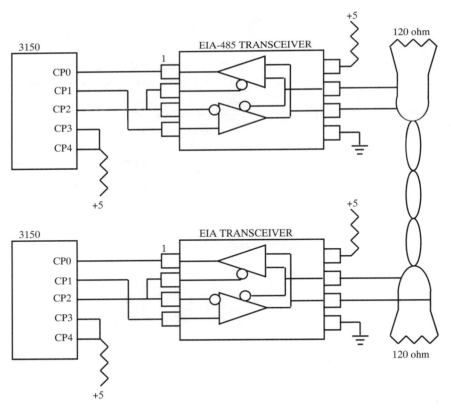

Figure C.18
EIA-485 twisted pair interface (used with single-ended mode)

C.3 ECHELON NEURON CHIP

Table C.4
Single-ended and differential network data rates

Network Bit Rate (kbit/s)	Minimum Input Clock (MHz)	Maximum Input Clock (MHz)
1,250	10.0	10.0
625	5.0	10.0
312.5	2.5	10.0
156.3	1.25	10.0
78.1	0.625	10.0
39.1	0.625	10.0
19.5	0.625	10.0
9.8	0.625	10.0
4.9	0.625	5.0

to V_{SS}) input and output buffers on pins CP0 and CP1. Single-ended and differential modes use differential Manchester encoding which is a widely used and reliable format for transmitting data over various media. The available network bit rates for single-ended and differential modes are given in Table C.4, as a function of the Neuron chip's input clock rate.

When put in direct single-ended mode the Neuron chip can interface to an external EIA-485 transceiver IC. While not having the performance of a transformer coupled circuit, this can offer a more economical solution. EIA-485 chips can be found in bipolar and CMOS versions. CMOS use less power but are usually more expensive. The communications port must be put in single-ended mode for implementing an EIA-485 network. EIA-485 specifies 32 drivers and 32 receivers on a single twisted pair of wires. Many wire types can be used with this approach, but collision detection may be difficult to implement.

In the differential mode (Figure C.19), the Neuron chip's built-in transceiver is able to differentially drive and sense a twisted-pair transmission line with external passive components. Differential mode is similar in most respects to single-ended mode; the key difference is that the driver/receiver circuitry are configured for differential line transmission. Data output pins CP2 and CP3 are driven to opposite states during transmission and put in a high-impedance (undriven state) when not transmitting. The differential receiver circuitry on pins CP0 and CP1 has a selectable low-pass filter with four selectable values of transient pulse (noise) suppression. The selectable hysteresis and glitch filter permit optimizing receiver performance to line conditions.

Figure C.20 below shows a typical packet waveform in differential mode. Note that the packet format is identical to that of the single-ended mode; the coding and jitter tolerance also apply identically.

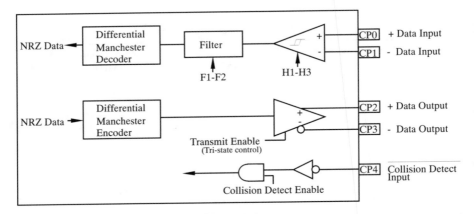

Figure C.19
Differential mode

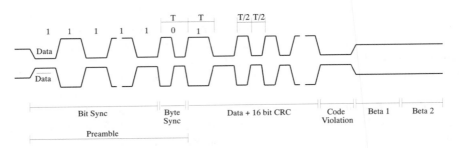

Figure C.20
Differential mode data format

The drivers can source 35 mA of current at 4.2 V and can sink 35 mA of current at 0.8 V ($V_{DD} = 5$ V). The programmable hysterests values shown in Table C.5 correspond to the differential threshold of the receiver. It is recommended that the signal level at the receiver be kept two to three times higher than the programmed threshold.

In single-ended mode (Figure C.21), the communications port encodes transmitted data and decodes received data using differential Manchester coding (also known as bi-phase space coding). This scheme guarantees a transition at the beginning of every bit period for the purpose of synchronizing the receiver clock. Zero/one data is indicated by the presence of absence of a second transition halfway between clock transitions. A mid-cell transition indicates a0. Lack of a mid-cell transition indicates a1. Differential Manchester coding is polarity-insensitive. Thus, reversal of polarity in the communication link will not affect data reception.

C.3 ECHELON NEURON CHIP

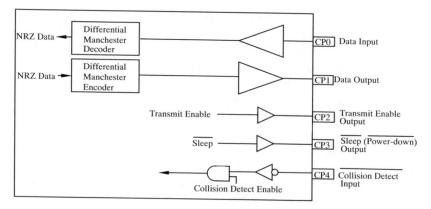

Figure C.21a
Single-ended mode configuration

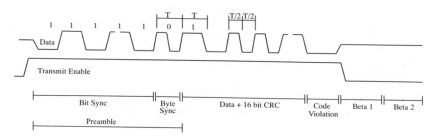

Figure C.21b
Single-ended mode data format

The transmitter transmits a preamble at the beginning of a packet to allow the other nodes to synchronize their receiver clocks. The preamble consists of a series of differential Manchester 1s. Its duration is at least six bits long and is selectable by the user.

The preamble ends with a single byte-sync bit, which marks the start of byte boundaries at the bit following. The byte-sync bit is a differential Manchester 0.

The Neuron chip terminates the packet by forcing a differential Manchester code violation. After sending the last CRC bit, it holds the data output transitionless for 2-1/2 bit times. The Transmit Enable pin on CP2 is then driven low, indicating the end of transmission. For a Neuron chip, the time to format and begin to send a 12 byte message is from 2.8 to 44.8 ms depending on the input clock rate (10 MHz–625 kHz).

As an option, the Neuron chip accepts an active-low collision detect input from the transceiver. If collision detection is enabled and CP4 goes low for at least one system clock period (200 ns with a 10 MHz clock) during transmission, the Neuron chip is signalled that a collision has or is occurring and that the message must be resent. The node then attempts to reaccess the channel.

If the node does not use collision detection, then the only way it can determine that a message has not been received is to request an acknowledgement. When acknowledged service is used, the retry timer is set to allow sufficient time for a message to be sent and acknowledged (typically 48 to 96 ms at 1.25 Mbps when there are no routers in the transmission path). If the retry timer times out, the node attempts to reaccess the channel. The benefit of using collision detection is that the node does not have to wait for the retry timer to time-out before attempting to resend the message, because the node detects the collision when it sends the packet.

In Figure C.22 Beta 1 time is the idle period after a packet has been sent and both priority (P) and non-priority slots are defined by the Beta 2 time. Nodes listen to the network prior to transmitting a packet. This prevents nodes from transmitting packets on top of each other except when the packets are initiated at nearly the same time. In addition, nodes randomize the time before they start transmitting on the network. When the network is idle, all nodes randomize over 16 slots. When the estimated network load increases, nodes start randomizing over more slots in order to lower the probability of a collision. The number of randomizing slots (R) varies from 16 up to 1008, based on n, the estimated channel backlog (the number of slots is $n \times 16$, where n has a range of 1 to 63).

In order for the receiver to detect the edge transitions, two windows are set up for each bit period T. The first window is set at $T/2$ and determines if a 0 is being received. The second window is at T and it defines a 1. This transition then sets up the next two windows ($T/2$ and T). If no transition occurs, a Manchester code violation is detected and the packet is assumed to have ended.

Transformer coupled twisted pair transceivers are designed to communicate between multiple remote nodes, and incorporate circuitry for common mode rejection, collision detection, line isolation, power-down protection, etc. When using transformers, collision detection can be supported over specified wire

Figure C.22
Packet timing

C.3 ECHELON NEURON CHIP

types and high common mode isolation and noise immunity is achieved. Properly designed transformer-coupled circuits can support up to 64 nodes on a single twisted pair of wires.

In special situations it is desirable for the Neuron chip to provide the packet data in an unencoded format and without a preamble. In this case, an intelligent transmitter accepts the unencoded data and does its own formatting and preamble insertion. The intelligent receiver then detects and strips off the preamble and formatting and returns the decoded data to the Neuron chip. Such an intelligent transceiver contains its own input and output data buffers, intelligent control functions, and provides handshaking signals to properly pass the data back and forth between the Neuron chip and the transceiver. In addition, there are many features that can be defined by and incorporated into a special-purpose transceiver.

References

Allik, K., Dunne, S., and Mulder, R. (1990) ArcoNet: A proposal for a standard network for communication and control in real-time performance. *Leonardo*, **23**(1), pp. 91–97.

Atkin, B.L. (1988) Progress towards intelligent buildings. *Intelligent Buildings*, B. Atkin (Ed.) UNICOM: Applied Information Technology Reports, 1988, pp. 1–7.

Baddoo, G.J.A., and Martin, D.P. (1991) Development and characteristics of PAKNET. *IEE Colloquium on Cordless Computing*, London, 1991, pp. 6.1–6.6.

Baxtor, T. (1988) Video distribution in the Eureka-IHS Network. *IEEE Transactions on Consumer Electronics*, **34**(3), pp. 736–743.

Bernaden, J.A., and Williams, A.F. (1989) *Open Protocols*. Fairmount Press, Georgia, USA.

Booth, P. (1990) *An Introduction to Human–Computer Interaction*. Lawrence Erlbaum Associates, London, UK.

Bowden, R. (1991) *The HART protocol*. Rosemount Ltd., Heath Place, Bognor Regis, West Sussex, UK.

Burton, P., Stone, F.W., and Kirk, M. (1988) *Fieldbus based on MIL-STD-1553B*. Proposal to ISA SP-50 Committee, October 17, 1988. ERA Technology Ltd., Leatherhead, Surrey, UK (Presented at *Industrial Fieldbus Seminar*, University of Warwick, 3 and 4 November, 1988).

Butler, A.D. (1991) The impact of home bus standards on consumer product design: addressing the challenge for a better user interface. *IEEE Transactions on Consumer Electronics*, **37**(2), pp. 163–167.

Cadeira, C., Siebert, M., and Thomesse, J.P. (1993) Scheduling on fieldbus with smart transducers/transmitters. *Proc. Int. Symp. on Intelligent Instrumentation for Remote and On-Site Measurement*, IMEKO, Brussels, May 1993, pp. 259–265.

Camp, C. (1992) Standards Talk: CENELEC and European standards harmonisation. *IT Technical Journal*, May–June, pp. 126–133.

Carter, A. and Pratt, A.R. (1993) Vehicle traffic detectors. *Measurement and Control* **26**(2), 1993, pp. 36–44.

Catling, I., and McQueen, B. (1991) Road transport informatics in Europe — major programs and demonstrations. *IEEE Transactions on Vehicular Technology*, **40**(1), pp. 1332–140.

Cawkell, A.E. (1991) IT in the home. Part 1. Expectations. *Critique* (Association for Information Management), **4**(4), pp. 1–12.

Cawkell, A.E. (1992) IT in the home. Part 2. Fulfilment? *Critique* (Association for Information Management), **4**, (5), pp. 1–12.

CEN (1991) *Safety of machines*. European Committee of Standardization (CEN). EN 292-1/2: 1992.

Clapp, M.D., and Churches, K. (1993) The future of building services management and control systems. *GEC Review*, **8**, (2), pp. 93–101.

Cohen, J. (1994) Automatic identification and data collection systems. McGraw Hill, 1994.

Collins, G.B. (1968) A survey of digital instrumentation and computer interface methods and developments. Proc. Conf. on Industrial Measurement Techniques for On-Line Computer. *IEE Conference Publication No. 43*, June 1968, pp. 1–8, 60–73.

Conner, D. (1988) The deterministic character of Arcnet proves ideal for the factory floor. *EDN*, September 15th, pp. 101–110.

Coschieri, J.C., and Troian, P. (1989) PC protocol analyser for the Eureka Integrated Home System project. *IEEE Transactions on Consumer Electronics*, **35**(3), pp. 552–556.

Cruickshank, A.M., and Kennett, M.J. (1989) *The development and flight demonstration of an optical MIL-STD-1553B data bus*. ERA Report No. 89-0591, ERA Technology Ltd., pp. 3.1.1–3.1.9.

Daniel, R.W., and Sharkey, P.M. (1990) Transputer control of a Puma 560 robot via the virtual bus. *IEE Proceedings*, **137D**, (4), pp. 245–252.

Denyer, P.B., and Renshaw, D. (1985) *VLSI Signal Processing: a bit-serial approach*. Addison-Wesley.

Dettmer, R. (1992) Meter power. *IEEE Review*, June, pp. 237–240.

Dheere, R.F.B.M. (1988) *Universal Computer Interfaces*. European Patent Office Applied Technology Series v. 11, Pergamon Press.

Di Giacano, J. (1990) *Digital Bus Handbook*, McGraw Hill.

REFERENCES

Dickson, D.M. (1991) A transceiver for the EIA Consumer Electronic Bus. *IEEE Transactions on Consumer Electronics*, **37**, (2), pp. 116-121.

Dooley, T. (1991) The road ahead for remote metering. *Electrical Review*, **224**, (5), pp. 14-15.

Dorey, H.A. (1977) The remote control of instrumentation. *Proceedings of the Conference on Programmable Instruments*, National Physical Laboratory, UK, November 1977, pp. 97-104.

Dostert, K.M. (1992) A novel frequency hopping spread spectrum scheme for reliable power line communications. *IEEE Second Int. Symp. on Spread Spectrum Techniques and Applications*, Japan, 1992, pp. 183-186.

Dostert, K.M. (1988) Multi-channel transmission of measurement and control data over indoor power lines using spread spectrum techniques. *IMEKO*, 1988, pp. 151-158.

Douligeris, C. (1993) Intelligent home systems. *IEEE Communications Magazine*, October, pp. 52-61.

Dwyer, J., and Ioannou, A. (1989) MAP and TOP: *Advanced Manufacturing Communications*, Kogan Page, London, UK.

ERA Report (1990) *The VXIbus: a review of its capabilities and compatible products*. ERA Report No. 90-0148. ERA Technology Ltd., Leatherhead, Surrey, UK.

Evans, G. (1991a) The EIA consumer twisted pair network. *IEEE Transactions on Consumer Electronics*, **37**, (2), pp. 101-107.

Evans, G. (1991b) Solving home automation problems using artificial intelligence techniques. *IEEE Transactions on Consumer Electronics*, **37**, (3), pp. 395-400.

Farrell, J. (1990) The economics of standardization: a guide for non-economics. In *An analysis of the Information Technology Standardization Process*, D.L. Berg and H. Schumny (Eds). Elsevier Science Publishers, B.V. (North Holland).

Fensome, D.A. (1990) The transputer — a protoyping tool for systems. *Computing and Control Engineering Journal*, January, pp. 41-45.

Fey, S. (1991) Infinity — the intelligent building has arrived. *Technology in Action*, BICC, January, pp. 28-30.

Figler, A.A., and Stead, S.W. (1990) The medical information bus. *Biomedical Instrumentation and Technology*. March/April, pp. 101-111.

Finley, M.R., Karakura, A., and Nbogni, R. (1991) Survey of intelligent building concepts. *IEEE Communications Magazine*, April, pp. 18-23.

Fisher, G. and Sullivan, T. (1993) Towards the integrated road transport environment. *Measurement and Control*, **26**, February, pp. 14-18.

Fitzpatrick, K.S., and Markwalter, E.B (1989) CEbus network layer description. *IEEE Transactions on Consumer Electronics*, **35**, (3), pp. 571–576.

Folts, H.C. (1980) X.25 transaction oriented features — datagram and fast select. *IEEE Transactions on Communications*, **28**, pp. 496–500.

Franklin, D.F., and Ostler, D.V. (1989) The P1073 Medical Information Bus. *IEEE MICRO*, October, pp. 52–60.

Freeburg, T.A. (1991) Enabling technologies for wireless in-building network communications — four technical challenges, four solutions. *IEEE Communication Magazine*, April, pp. 58–64.

Freer, J. (1987) *System design with advanced microprocessors*. Pitman, London, UK.

Gao, Yang, and Durrant-Whyte, H.F. (1991) A tranputer-based sensing network for process plant monitoring. *Application of Transputers 3* (V. 1) IOS Press, pp. 180–185.

Gardner, R.M., Hawley, W.L., East, T.D., Oniki, T.A., and Young, H.F.W. (1992) Real time data acquisition: recommendations for the Medical Information Bus (MIB). *Int. J. of Clinical Monitoring and Computing*, **8**, pp. 251–258.

Gater, C. (1987) *Fault tolerant distributed measurement systems*. PhD Thesis. University of Edinburgh.

Gershon, R., Propp, D., and Propp, M. (1991) A token passing network for powerline communications. *IEEE Transactions on Consumer Electronics*, **37**, (2), pp. 129–134.

Graves, R. (1989) *In the 'light' of experience*. ERA Report No. 89-0591, ERA Technology Ltd., pp. 6.1.1–6.1.12.

Gray, P. (1991) *Open Systems: A Business Strategy for the 1990's*. McGraw Hill.

Griffiths, J.M. (1990) *ISDN Explained*. Wiley, UK.

Groak, S. (1992) The Idea of Building. E&FN Spon., 1992.

Gustavson, D.B. (1992) The Scalable Coherent Interface and related standards project. *IEEE Micro*, February, pp. 10–22.

Hammond, P.H., and King, P.J. (1980) A prospect for industrial control. *Electronics and Power*, January, pp. 38–45.

Hamabe, R. et al. (1988a) System design of the super home bus system (S-HBS) for apartment buildings. *IEEE Transactions on Consumer Electronics*, **34**, (2), pp. 327–333.

Hamabe, R. et al. (1988b) A protocol example of Super Home Bus Systems. *IEEE Transactions on Consumer Electronics*, **34**(3), pp. 686–693.

REFERENCES

Harmer, A.L. (1991) European research on advanced sensors. *Sensors: Technology, Systems and Applications*. K.T.V. Grattan (Ed.). Adam Hilger, Bristol, UK.

Hatamien, M., and Bowen, E.G. (1983) Homenet — a broadband voice/data/video network on CATV systems. *AT & T Tech. J.*, **64**, (2), pp. 347–367.

HMSO (1990) *Energy efficiency in domestic appliances*. Department of Energy Report, No. 13, (in the Energy Efficiency Services). HMSO, UK.

Hofman, J. (1991) The Consumer Electronic Bus infrared system. *IEEE Transactions on Consumer Electronics*, **37**, (2), pp. 122–128.

Hoshikuki, A., Yamamoto, M., Ishii, S., Kohno, R., and Imai, H. (1992) Implementation of an industrial R/C system using a hybrid DS/FH spread spectrum technique. *IEEE Second Int. Symp. on Spread Spectrum Techniques and Applications*, Japan, pp. 179–182.

IEC (1992) Draft *Functional safety of electrical/electronic/programmable systems*. Generic Aspects. Part 1: General Requirements, IEC Draft Standards.

ISIbus Brochure (1990) Senter for Industriforskning, Forskningsveien 1, P.O. Box 124 Blindern, 0314 Oslo, Norway.

Inmos (1988) *Tranputer Reference Manual*. Prentice Hall.

Iyengar, S.S., Jayasimha, D.N., and Nadig, D. (1994) A versatile architecture for the distributed sensor integration problem. *IEEE Transactions on Computers*, **43**, (2), pp. 175–185.

Jordan, J.R., Gater, C., and Mackie, R.D.L. (1989) Fault-tolerant loops for distributed measurement systems. *IEE Proceedings*, **136E**, (6), pp. 485–489.

Jordan, J.R., Lytollis, S., and Kent, D.W. (1992) A fibre optically extended fieldbus. *Measurement Sci. Technol.*, **3**, pp. 902–908.

Judge, P. (1988) Open Systems: The basic guide to OSI and its implementation. *Computer Weekly Publications*.

Jurgens, R.K. (1991) Smart cars and highways go global. *IEEE Spectrum*, May, pp. 26–36.

Jurgens, R.K., and Perry, T.S. (1985) The High-Tech Home. *IEEE Spectrum*, **22**, (5), pp. 35–112.

Kalani, G. (1988) *Microprocessor based distributed control systems*. Prentice Hall.

Khawand, J. et al. (1991) Common Application Language (CAL) and its integration into a home automation system. *IEEE Transactions on Consumer Electronics*, **37**, (2), pp. 157–162.

Khawand, J. et al. (1992) A physical layer implementation for a twisted pair home automated system. *IEEE Transactions on Consumer Electronics*, **38**, (3), pp. 530–536.

Kirk, M., and Wood, G. (1994) *The Executive Guide to Fieldbus*. Department of Trade and Industry Booklet, UK.

Kirrman, H.D. (1987) Fault tolerance in process control. *IEEE Micro*, October, pp. 27-50.

Knoll, S., and Colo, S. (1992) *RS-485 cabling guideline for the COM 20020 universal local area network controller and experimental procedure for verification of RS-485 cabling guidelines*. Technical Note 7-5, Standard Microsystems Corporation, Component Products Division, 35 Marcus Blvd., Hauppauge, New York 11788, USA.

Leveson, N.G. (1991) Software safety in embedded computer systems. *Communications of the ACM*, **34**, (2), pp. 34-36.

Lin, T.P. (1990) A Muiltifunction ISDN Home Communication System. *IEEE Transactions on Consumer Electronics*, **36**, (4), pp. 892-896.

Littlewood, B., and Strigini, L. (1992) The Risks of Software. *Scientific American*, November, pp. 38-43.

Liu, C., and Leyland, J. (1973) Scheduling algorithms for multiprogramming in a hard real-time environment. *Journal of the ACM*, **20**, (1), pp. 46-61.

Loughry, D.C. (1978) IEEE Standard 488 and microprocessor synergism. *Proc. IEEE*, **66**, (2), pp. 162-172.

Luo, R.C., and Kay, M.G. (1989) Multisensor integration and fusion in intelligent systems. *IEEE Transactions on Systems, Man and Cybernetics*, **19**, pp. 901-927.

Lytollis, S., Jordan, J.R., and Kelly, R.G. (1989) *A hybrid twisted pair/optical fieldbus based on MIL-STD-1553B*. ERA Report N.89-0591. ERA Technology Ltd, pp. 3.2.1-3.2.10.

Madron, T.W. (1989) *LANS - Applications of IEEE/ANSI 802 Standards*. Wiley, 1989.

Markwalter, B.E. *et al.* (1991) Design influences for the CEbus automation protocol. *IEEE Transactions on Consumer Electronics*, **37**, (2), pp. 145-153.

Marshall, A., and Spracklen, C.T. (1990) Network architectures for a direct sequence spread spectrum LAN. *Proc. IEEE Int. Symp. on Spread Spectrum Techniques and Applications*, London, pp. 176-180.

Marzullo, K. (1990) Tolerating failures of continuous-valued sensors. *ACM Transactions on Computer Systems*, **8**, (4), pp. 284-304.

Mitchell, D.A.P., Thompson, J.A., Mansons, G.A., and Brookes, G.R. (1990) *Inside the Transputer*. Blackwell Scientific Publications.

Morgan, E. (1987) *Through MAP to CIM*, Department of Trade and Industry, London, UK.

Muratia, M. *et al.* (1983) A proposal for standardization of the home bus system for home automation. *IEEE Transactions on Consumer Electronics*, **29**, (4), pp. 524–529.

Neve, B.D. (1990) Progress on radio fieldbus. *Measurement and Control*, **23**, (2), pp. 14–19.

Northcott, J. (1991) *Britain in 2010*. Policy Studies Institute.

Packer, J.S. (1990) Patient care using closed-loop computer control. *Computing and Control Engineering Journal*, January, pp. 23–28.

Papovic, D., and Bhatkar, V.P. (1990) *Distributed computer control for industrial automation*. Marcel Dekker.

Parkes, A.M. (1992) Car instrumentation in the future. *Measurement and Control*, **25**, November, pp. 261–268.

Parnas, D.L., van Schouwen, A.J., and Kwan, S.P. (1990) Evaluation of safety critical software. *Communications of the ACM*, **33**, (6), pp. 636–648.

Pellerin, D., Brissand, M., and Grange, G. (1990) Single chip microcomputer based intelligent sensors for home automation network. *IEE Proceedings*, **137**, E, (5), pp. 359–365.

Phillips, T. (1994) Welcome to the computerised home. *The Guardian*, March 10th, p. 19.

Pimentel, J.R. (1989) Communication architectures for fieldbus networks. *Control Engineering*, October, pp. 74–78.

Potter, B., Sinclair, J., and Till, D. (1991) *An introduction to formal specification and Z*. Prentice Hall.

Powell, J.A. (1988) Towards the independent environment for intelligent buildings. In *Intelligent Buildings*. B. Atkin (Ed.). UNICOM: Applied Information Technology Reports, pp. 234–251.

Rabbie, H.M. (1992) Distributed processing using local area networks. *Assembly Automation*, **12**, (1), pp. 14–19.

Rausch, R. (1989) *Control of the large European LEP and SPS accelerators based on the 1553 fieldbus*. ERA Report No. 89-0591, ERA Technology Ltd, pp. 2.1.1–2.1.12.

Robertson, B., Chopping, M., Zielinski, K., and Milway, D. (1990) The Metobridge — an application of transputers in transparent bridging. *Application of Transputers 2*, IOS Press, pp. 303–310.

Rogers, R. (1991) Architecture: A Modern View. Thames and Hudson, 1991.

Rothstein, J. (1992) *MIDI: A Comprehensive Introduction*. Oxford University Press.

Sakamura, K.A. (1993) Toward a world filled with computers. *IEEE Micro*, October, pp. 6-11.

Saund, T.S., Comley, V.E., and Hill, P.C.J. (1990) Data communication over power circuits using direct sequence spread spectrum modulation. *IEEE Spread Spectrum Symposium*, pp. 25-29.

Ship Star (1994) *ISP Product Bulletin*. Ship Star Associates Inc., 36 Woodhill Drive, Suite 19, Newark, DE 19711-7017, USA.

Shladover (1991) *IEEE Transactions on Vehicular Technology*, **40**, (1), p. 116.

Shneiderman, B. (1993) Beyond intelligent machines. *IEEE Software*, January, pp. 100-103.

Shneiderman, B. (1992) *Designing the user interface: strategies for effective human-computer interaction*. Addison-Wesley.

Smythe, C. (1993) ISO 8802/3 local area networks. *Electronics and Communication Engineering Journal*, February, pp. 25-33.

Stallings, W. (1991) *Data and Computer Communications*. Macmillan.

Stone, F.W.C. (1989) *1553B fieldbus*. ERA Report No 89-0591, ERA Technology Ltd, pp. 2.2.1-2.2.13

Sumner, L. (1991) Standards, product liability and the consumer. *Engineering Management Journal*, February, pp. 19-25.

Surrat, J.M. (1991) Integration of CEBus with utility load management and automatic meter reading. *IEEE Transactions on Consumer Electronics*, **37**, (3), pp. 406-412.

Tester, J.W., Wood, D.O., and Ferrari, N.A. (1991) *Energy and the Environment in the 21st Century*. The MIT Press, 1991.

Tobagi, F., and Kleinrock, L. (1976) Pocket switching in radio channels: Part III — Polling and dynamic split-channel reservation multiple access. *IEEE Transactions Communications*, **24**, pp. 832-845.

Tritton, J.A. (1988) Interactive home systems — an overview. *IEEE Transactions on Consumer Electronics*, **34**, (3), pp. 693-699.

Tuinenburg, H.A. (1990) Conformance testing of IT implementations. In *An analysis of the Information Technology Standardization Process*, J.L. Berg and H. Schumny (Eds.). Elsevier Science Publishers, B.V. (North-Holland), pp. 73-83.

UK HSE (1987) *Programmable electronic systems in safety related applications*. UK Health and Safety Executive.

Von Tomkewitsch, R. (1991) Dynamic route guidance and interactive transport management with ALI-SCOUT. *IEEE Transactions on Vehicular Technology*, **40**, (1), pp. 45-50.

REFERENCES

Wacks, P.K. (1991) Utility load management using home automation. *IEEE Transactions on Consumer Electronics*, **37**, (2), pp. 168–174.

Weiser, M. (1991) The computer for the 21st Century. Scientific American **265**(3), 1991, pp. 66–75.

Welsh, D. (1988) *Codes and Cryptography*. Oxford University Press.

Wilkins, C. (1990) Applying computers in the National Health Service. *Computing and Control Engineering Journal*, January, pp. 29–34.

Winick, S.J. (1991) The RF medium in the home — the move to spread spectrum. *IEEE Transactions on Consumer Electronics*, **27**, (2), pp. 108–115.

Wood, G. (1990) Fieldbus standardization in 1990. *Measurement and Control*, **23**, (5), pp. 135–139.

Index

AFNOR, 177
AGV, 65, 138
Analogue signalling, 32
ANSI, 56, 177
ARINC 629, 12, 113, 152, 177
Arcnet, 6, 73, 141, 177, 179
ASIC, 41, 79

Bitbus, 73, 75
Bit-oriented, 5
Bridge, 67
British Standards Institution, 177

CAMAC, 10, 88
CAN, 6, 74, 117, 170, 193
CE bus, 40, 108, 162
CEN, 56
CENELEC, 56
Characteristic impedance, 18
Civil aviation, 113
Closed loop, 78
Coaxial cable, 17
Command-response, 48
CSMA, 50

Digital codes, 35
DIN, 177
Domestic buildings, 97

Echelon LonWorks, 74, 108, 165
Echelon Neuron Chip, 210
EIA, 178
EIA RS Standards, 42
EPA, 47
ERA, 113
Error, 36

Factory automation, 64
Fibre optic, 20

FIP, 72, 157
Flexible manufacturing, 136

Gateway, 67
GPIB, 89
Ground loops, 18

HART, 81, 134, 159
HDLC, 5, 73, 150
Home automation, 101
Home bus system, 108, 162
Honeywell, 62

I^2C, 14, 125
IEC, 55, 178
IEC Fieldbus, 9, 50, 68, 75, 174
IEEE, 178
IEEE 488, 89
IEEE 802, 5
IEEE 1118, 73, 150
IEEE P1073, 11, 93, 168
IEEE P1157, 91
IEEE P1394,
Inductive loops, 119
Infrared beacons, 119
Intelligent activator, 80
Intelligent buildings
    domestic, 97
    commercial, 104
Intelligent instrumentation, 48
Intelligent site, 107
Intelligent transducers, 76
Intelligent vehicle-highway system, 118
Interfaces
    Man-machine, 45
    Machine-machine, 45
Integrated circuits
    Arcnet bus controller, 179

CAN bus controller, 193
Neuron chip, 210
I^{2}C controller, 127
Intrinsic safety, 32
ISA, 178
ISO, 55, 178

Junction box, 19

Laboratory automation, 87
Link active schedule, 51
Link master, 51
Low power, 20, 27, 28, 82, 100, 113

Machine automation protocol, 47, 66
Manchester code, 35
Master-slave, 48
Media access, 48
Medical instrumentation, 91
MIB, 11, 93
MIL-STD-1553, 13, 109, 145
MIL-STD-1773, 13, 109, 145
Military systems, 109,
Mini-MAP, 47
MOD 300, 3
Multidrop, 1
Multistar topology, 26
Musical Instrumentation Digital Interface, 131

Network management, 70
NRZ, 35
NRZ1, 150

Open System Interconnect, 46

Packet, 30
PDU model, 48
Power line, 39,
Power over the bus, 81
Process control, 62
PROFIBUS, 72, 154
Propagation delay, 19
PROWAY, 62

Quality of Service, 69

Radio Data System, 119

Radio fieldbus, 32
Radio link, 27
Redundancy, 84
Reflective star, 25
Remote metering, 100
Remote terminal, 146
Ring topology, 22, 51
Road automation, 118
Router, 67
RS 485, 42
RZ, 35

SAE J1850, 115
Safe systems, 83
SDLC, 5, 73, 150
Software reliability, 85
Spread spectrum, 38
Standard document, 57
Standards, 52
Standards organisations, 54
Synchronisation, 146
System integrator, 136

Telemetry bands, 28
Telemetry low power, 31
THORP, 3
Timers, 143
Token passing, 49
Transmissive star, 25
Transputer, 121
Twisted pair, 17
Twists per metre, 17

ULSIC, 14

VAN, 115
Vehicles, 113
VME, 89
VXIbus, 90

Wireless LAN, 29

X25, 29

Y splitters, 23

Zener barrier, 34

Books are to be returned on or before the last date below.

Books are to be returned on or before the last date below.

24 JUL 2004

13 NOV 2004

02 APR 2005

LIBREX—